QUALITY-ORIENTED DESIGN OF BUSINESS PROCESSES

LANCHESTER LIBRARY, Coventry University

Much Park Street   Coventry CV1 2HF   TEL: 01203 838292

QUALITY-ORIENTED DESIGN OF BUSINESS PROCESSES

by

Kai Mertins
Roland Jochem
Fraunhofer Institute IPK Berlin

Kluwer Academic Publishers
Boston/Dordrecht/London

Distributors for North, Central and South America:
Kluwer Academic Publishers
101 Philip Drive
Assinippi Park
Norwell, Massachusetts 02061 USA
Telephone (781) 871-6600
Fax (781) 871-6528
E-Mail <kluwer@wkap.com>

Distributors for all other countries:
Kluwer Academic Publishers Group
Distribution Centre
Post Office Box 322
3300 AH Dordrecht, THE NETHERLANDS
Telephone 31 78 6392 392
Fax 31 78 6546 474
E-Mail <orderdept@wkap.nl>

   Electronic Services <http://www.wkap.nl>

Library of Congress Cataloging-in-Publication Data

Mertins, K. (Kai), 1947 -
  Quality-oriented design of business processes / Kai Mertins, Roland Jochem.
      p. cm.
  Includes bibliographical references and index.
  ISBN 0-7923-8484-9
  1. Production management. 2. Quality control. 3. Business--Graphic methods.   I. Jochem, R. (Roland), 1962-   . II. Title.
TS155.M44   1999
658.5--dc21   99-29500
                                                                               CIP

Copyright © 1999 by Kluwer Academic Publishers.

All rights reserved. No part of this publication may be reproduced, stored in a retrieval system or transmitted in any form or by any means, mechanical, photo-copying, recording, or otherwise, without the prior written permission of the publisher, Kluwer Academic Publishers, 101 Philip Drive, Assinippi Park, Norwell, Massachusetts 02061

Printed on acid-free paper.

Printed in the United States of America

Contents

Chapter 1: Business Process Design and Quality Management   1
    1. THE COMPANY AS A SYSTEM   1
    2. MODEL   3
    3. METHOD AND MODELING LANGUAGE   5
    4. BUSINESS PROCESS   6
    5. QUALITY MANAGEMENT   7
    6. BUSINESS PROCESS MODEL   7
    7. BUSINESS PROCESS DESIGN   7

Chapter 2: Requirements of Business Process Design   9
    1. INTRODUCTION   9
    2. REQUIREMENTS OF ORGANIZATIONAL DEVELOPMENT   9
    3. REQUIREMENTS OF QUALITY MANAGEMENT   11
    4. REQUIREMENTS OF INFORMATION SYSTEMS PLANNING   13
    5. REQUIREMENTS OF CONTROLLING AND COST ACCOUNTING   14
    6. SUMMARY AND NEED FOR ACTION   15

Chapter 3: Quality-Oriented Design of Business Processes   17
    1. INTRODUCTION   17
    2. MODELING LANGUAGE TO DESCRIBE BUSINESS PROCESSES AND QUALITY ASPECTS   18
       2.1 Basic Constructs 18
       2.2 Views.. 23
       2.3 Process Modeling... 25

      2.4 EXPRESS for the Formal Description of the Modeling
          Language 29
  3. INTEGRATION OF QUALITY MANAGEMENT   36
      3.1 Process-Oriented Integration of Quality Management 36
      3.2 Descriptive Rules for Quality Management 38
      3.3 QM Elements as Process Modules 43
      3.4 Support of the Corporate QM System through Control
          Circuit Models and Process Models to Derive Indicators 46
  4. REFERENCE MODELS AND MODEL LIBRARIES   49
      4.1 Introduction 49
      4.2 Reference Model 'Order Throughput' 50
      4.3 Model Library Quality Management 61

Chapter 4: Modeling Rules for Quality-Oriented Design of Business Processes   65
  1. GOALS AND REQUIREMENTS OF MODELING RULES   65
      1.1 General Requirements of Modeling Rules 66
      1.2 QM-typical Requirements of Modeling Rules 66
  2. APPROACHES TO QUALITY-ORIENTED MODELING   67
      2.1 Modeling Steps 67
      2.2 Ways to Quality-Oriented Modeling 69
  3. GOAL FINDING AND SYSTEM DELIMITATION   72
      3.1 Goals 72
      3.2 Determination of the Limits of a Model...72
      3.3 Determination of the Desired Level of Detail....................................73
  4. MODEL CREATION   74
      4.1 Goals and Desired Results 74
      4.2 Identification of Objects and Object Classes to be Modeled............75
      4.3 Creation of Process Models 75
      4.4 Creating QM-Related Process Models 77
      4.5 Creation of Information Models 81
      4.6 Creating QM-Related Information Models..82
      4.7 Development of Specific Submodels 83
  5. MODEL EVALUATION   83
      5.1 Targets and Possibilities of a Model Evaluation...............................83
      5.2 Specification of Improvement Potentials in
          Quality Management..84
  6. MODEL MODIFICATION   86
      6.1 Targets and Desired Results 86
      6.2 Determining the Base Model 87
      6.3 Identifying the Modification Requirements of an
          Existing Model 87
      6.4 Executing the Necessary Operations..88

Chapter 5: Computer-Based Tool — 89
1. SITUATION, GOALS AND REQUIREMENTS — 89
2. TOOL CONCEPT — 90
3. THE ENTERPRISE MODEL IN THE TOOL — 91
   - 3.1 Modeling of Views — 91
   - 3.2 Parameters — 92
   - 3.3 Integration of STEP/EXPRESS — 93
4. STRUCTURE OF THE TOOL — 94
   - 4.1 Modular Structure — 94
   - 4.2 Hardware and Software Requirements — 94
   - 4.3 Data Model — 95
5. USER INTERFACE AND FUNCTIONALITY — 96
   - 5.1 Modeling Components — 96
   - 5.2 Class Editors — 96
   - 5.3 Components' Editors — 96
   - 5.4 Business Process Editor — 97
   - 5.5 Editing Attributes / Parameters — 98
   - 5.6 Navigation in the Model — 99
6. INTERFACES — 99
   - 6.1 Interfaces with MS®-WINDOWS™ Applications — 99
   - 6.2 Database Interfaces — 99
   - 6.3 Export / Import - Interfaces — 99
7. EVALUATION MECHANISMS — 100
   - 7.1 Predefined Evaluations — 100
   - 7.2 Evaluations that can be Parametrized — 101
   - 7.3 User-Specific Evaluations — 101
   - 7.4 Example of an Evaluation — 102
8. SUMMARY AND DIRECTIONS OF DEVELOPMENT — 103

Chapter 6: Model-Based Development of Quality Management Documents — 107
1. QM SYSTEMS AND QM DOCUMENTS — 107
2. METHODS OF DOCUMENT DEVELOPMENT — 108
   - 2.1 Principle of Element-Oriented QM Document Development — 108
   - 2.2 Program-Technical Support of the Element-Oriented Approach — 109
   - 2.3 Principle of Model-Based QM Document Development — 111
3. DESCRIPTION OF THE QM DOCUMENTS WITH IEM METHOD — 115
   - 3.1 Document Structure — 115
   - 3.2 Development of a QM Document — 116
4. AUTOMATIC GENERATION OF QM DOCUMENTS — 122
5. SUPPORT OF ISO 9000 CERTIFICATION AND IMPLEMENTATION — 123
   - 5.1 Supporting the Certification and Implementation Process — 123

	5.2 ISO 9000 Reference Models	126
	5.3 Descriptive Rules of IEM Reference Models	127
6.	BENEFITS OF MODEL-BASED QM DOCUMENT DEVELOPMENT	132

Chapter 7: Case Study 133

1.	TARGETS AND APPROACH	133
2.	DESCRIPTION OF THE COMPANY	135
3.	ACTUAL STATE ANALYSIS	136
	3.1 Process Structure of the Company	136
	3.2 Description of Time, Cost and Quality Requirements	141
	3.3 Development and Description of Weak Points	143
4.	TARGET CONCEPT	150
	4.1 Design	150
	4.2 Production Planning	151
	4.3 Design and Production of Manufacturing Devices	153
5.	IMPLEMENTATION OF QM SYSTEM	156
	5.1 Goals	156
	5.2 Approach	156
	5.3 Short Summary of Results	157
	5.4 Target Concept and a Plan of Measures	158
6.	FINAL RESULT OF THE CASE STUDY	160

Chapter 8: Standardization 161

1.	INTRODUCTION	161
2.	NATIONAL	161
	2.1 NAM 96.5.1 'Framework for a CIM System Integration'	162
	2.2 NAM 96.4.8 'Industrial Manufacturing Data Management'	162
	2.3 NAM 96.4.4 'Methodology, Tests of Conformity and Implementation'	164
	2.4 Standardization Committee 'Quality Management, Statistics and Certification Elements (NQSZ) - Subcommittee 1 'Quality Management''	164
	2.5 Standardization Committee 'Sachmerkmale' (NSM)	164
3.	EUROPEAN	165
4.	INTERNATIONAL	165
	4.1 ISO/TC 184/SC 5/WG 1 „Modeling and Architecture'	165
	4.2 ISO TC 184/SC 4/WG 5 „EXPRESS Language'	166
	4.3 ISO/TC 184/SC 4/WG 8 „Manufacturing Management Data' (MANDATE)	166

Chapter 9: Summary	167
References	169
Appendix A	175
Appendix B	187

Authors

Kai Mertins (Prof. Dr.-Ing.)

Born 1947. Qualified electromechanical engineer. Studied electrical engineering at the Hamburg Engineering College. Varied experience as an electrical engineer in industry in the fields of planning, projecting and commissioning. Studied industrial engineering at the Technical University (TU) Berlin. 1984 doctorate from TU Berlin, thesis on 'Control of Computerized Manufacturing Systems'. Since 1982 head of department, since 1988 director at the Fraunhofer Institute for Production Systems and Design Technology (IPK Berlin), responsible for the division Systems Planning, with the subsidiary departments Production Management, Production Planning and Production Control. Since 1998 Professor for Global Production Engineering at Technical University of Berlin.

Roland Jochem (Dipl.-Ing.)

Born 1962. Studied mechanical engineering at the Technical University (TU) Berlin. Experience in industry as a mechanical engineer in the projection and planning of assembly systems. Since 1988 member of the scientific staff, since 1991 group leader for Business Process Modeling at the Fraunhofer Institute for Production Systems and Design Technology (IPK Berlin), division Systems Planning.

Preface

Changing markets and innovative competitors force each company to study and improve its organization, its business processes and its technologies constantly. Whoever drops behind in these times loses market shares and endangers the long-term existence of the company.

It is necessary to realign the entire corporate planning and design along the value added chain to speed up the business processes. All relevant user views, such as quality, organization, information systems, costs (controlling) and affected departments, have to be incorporated and should be studied regarding their interaction [Joc95b, Joc95c].

When designing and executing processes to accept and employ a quality concept the decisive aspects are determined by the kind of information and by the active integration of all affected departments, divisions and staff members. On the other hand, the staff members must have a good knowledge of important business processes. The employees of quality planning should be 'process advisors', i.e., quality management should be incorporated into the business process. The quality-oriented design of business processes described in this book supports this integration.

The book is based on the results of the KCIM project 'Scientific Basis and Contribution to the Standardization of CIM Interfaces' [KCIM89a, KCIM89b, Mer91, Spu93a] that was funded by the German Federal Ministry for Research and Technology. On this basis, we developed the concept 'Quality-Oriented Design of Business Processes' enabling the integration of quality management into the design and planning of business processes.

Following a conceptual description of the subject, we describe the requirements necessary for a quality-oriented design of business processes. We then present a modeling method, called IEM (Integrated Enterprise Modelling), in line with the related standards, with a modeling language to display processes transparently, to derive suitable measures, to determine the suitable measures with the persons responsible and to document and constantly update these in quality manuals [Mer95a].

By way of creating transparency through models this results in a support of the complex design and optimization processes. The knowledge of processes and a common understanding of all participants are achieved through a description of business processes and quality-assuring measures in a model. This enables users to evaluate the quality. This means that business processes are a basis of applying quality standards, of quality documentation and of certification [Mer95b].

The application of the concept 'Quality-Oriented Design of Business Process' is illustrated with an example of a company. A computer-aided tool, called MOOGO, supporting the introduced modelling method IEM, enables the required incorporation of all persons participating in the design and optimization process and the model-based creation of QM documents. Finally, we describe the entry of the IEM-method into national, European and international standardization.

The work on the book was conducted during the QCIM project 'Quality Assurance through CIM', sub-project 'Quality-Oriented Enterprise Modeling'. It represents an approach to the model-based design of business processes and model-based quality documentation.

The reader is enabled to design his company's processes in a quality-oriented way. The modeling and planning steps are described in such a way that they can either be carried out alone or in cooperation with a consultant and a methodical handbook.

Acknowledgements

The authors thank the following persons for their contributions and support:

Chapter 4:   Quality-Oriented Design of Business Processes.
Dipl.-Inform. S. Beck; Dipl.-Ing. J. Dommel; Dipl.-Math. J. Hofmann;
Dipl.-Inform., Dipl.-Ing. F.-W. Jaekel; Dipl.-Ing. M. Schwermer;
Dipl.-Ing. F. Ihlenfeld; Dipl.-Ing. G. Windhoff ;
Dipl.-Ing. O. Krah; Dipl.-Ing. O. Schellberg.

Chapter 5:   Modeling Rules for the Quality-Oriented Design of Business Processes.
Dipl.-Ing. M. Schwermer; Dipl.-Ing. G. Windhoff.

Chapter 6:   Computer-Based Tool.
Dipl.-Inform., Dipl.-Ing. F.-W. Jaekel.

Chapter 7:   Model-Based Creation of Quality Management Documents.
Dipl.-Ing. J. Dommel.

Chapter 8:   Case Study.
Dipl.-Math. J. Hofmann ; Dipl.-Ing. F. Ihlenfeld; Dipl.-Ing. G. Windhoff;
Dipl.-Ing. O. Krah; Dipl.-Ing. O. Schellberg.

Chapter 1

Business Process Design and Quality Management

1. THE COMPANY AS A SYSTEM

The systems theory provides definitions for abstract views of any aspects of a system. A system is an arrangement of objects, i.e., elements that interact and that are marked off from the environment by the systems boundary [Rie78, Pat82]. The figures that cross the boundary in either direction are the system's input and output (Figure 1).

Systems technology distinguishes between different concepts that could be described as views on systems [Roh75]: the functional, the hierarchical and the structural concept.
- The functional concept focuses on the transformation of input units into output units. This transformation describes the function of the system. The functional concept describes the behavior of the system and its elements.
- The structural concept focuses on 'elements' and their connecting 'relations'. Additionally, the system is marked off from its environment.
- According to the hierarchical concept the system consists of several subsystems and belongs to a supra-system.

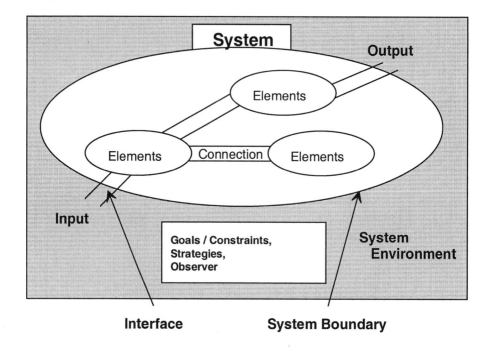

Figure 1: Model of a System

A complete description of systems should consider all three views. Applying the classification criteria of the systems theory a company could be characterized as a system in the following way:
- *artificial*: man-made,
- *dynamic*: the different elements react differently to each other,
- *real*: objective, observable,
- *open*: there are relations to the environment,
- *probabilistic*: the probability of predictions of the future behavior of the system is limited; however, the more people understand the system the better the predictions get.

Systems technological descriptions of a production have usually been taken from studies of machine tools and manufacturing systems [Wie87]. The operands form
- an information flow,
- a materials flow and
- an energy flow.

The input units 'information', 'energy' and 'materials', i.e., tools, auxiliary equipment and materials and raw materials, are transformed in the manufacturing process into the non-reversible output units 'finished items', 'waste' and 'heat' and into the semi-reversible output units 'information', 'tools' and 'auxiliary equipment and materials'. This study focuses on the functional aspects of the system.

Wieneke [Wie87] describes a manufacturing system as a system sealed by an envelope surface whose input units become elements of the system when they pass through the envelope surface. The elements include resources, auxiliary and operating materials, tools and auxiliary tools. They are characterized by certain attributes. The system's structure, process plans and the production control system are seen as relations. The functions of a manufacturing system are divided into manufacturing, assembling, testing, moving and resting. The dynamics of the system can be recorded through simulation in a model or in the real system. The hierarchical aspects are taken into account when analyzing functions. In total, the focus is on functional aspects, though.

Mertins [Mer85] divides manufacturing systems into the subsystems 'information system', 'energy system', 'materials flow system' with a 'storage', 'transportation' and 'handling system' and 'processing system'. This distinction focuses on structural and hierarchical aspects.

Wiener calls a company a techno-economic-social system. It is a dynamic entity whose parts (subsystems) are connected in such a way that no part is independent. The behavior of this entity is influenced by the interaction and by the emerging 'integrity' [Wir92].

We can conclude that the systems technological descriptions focus on operands and elements as well as on the processes that change them.

2. MODEL

A model is a system describing another, real system [Vie86]. Model creation requires suitable modeling methods. A method is an approach to solve certain tasks. A modeling method consists of constructs as well as of procedures describing how to use the constructs effectively when creating models (Figure 2). Constructs consist of the elements of a 'descriptive language' as well as the rules for the connection of elements. For example, a programming language consists of constructs that describe programs.

The term 'model' is characterized in detail by descriptive features, reduction features and pragmatic features [Schm85].

The *descriptive feature* characterizes the model as a reproduction of the real system. We distinguish between isomorphic, i.e., structurally equal, and homomorphic, i.e., structurally similar, models. An isomorphic model attempts to reproduce all elements and the connecting relations truly.

Homomorphic models display some elements and relations of a system as well as generalized features of the system.

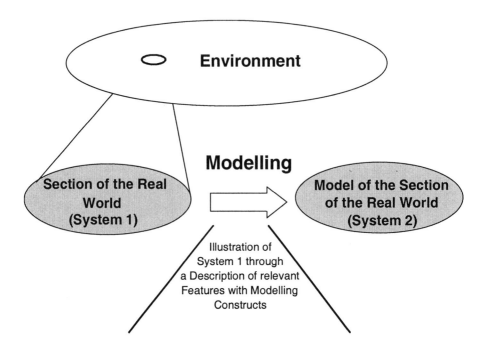

Figure 2: Reality and Model

Reduction features of models arise from the fact that modelers will only describe those elements and relations that seem to be relevant. Therefore, it is likely that different modelers working on the same system come up with different models.

The *pragmatic feature* describes the objective and the purpose of the model. A model should allow insight into the structure and the behavior of the depicted system. Therefore, a model is usually a simplified version of reality – according to the reduction features, and, if certain features need to be stressed, according to the descriptive feature. Depending on the modeling purpose we can distinguish descriptive, explanatory and decision models.

The purpose of *descriptive models* is to collect information on the condition of a system that is regarded as static. The modeler describes the system and decision moments. *Explanatory models* are to explain and forecast real aspects of a system, i.e., they should enable users to foresee the future condition of the system. *Decision models* should develop certain actions from pre-defined objectives, initial conditions and decision variables.

Models, just as systems, can be classified according to different criteria. Figure 3 illustrates possible criteria and their characteristics.

Model Classification	
Feature	**Characteristics**
Form of Presentation	iconic, analogusly, symbolically
Form of Illustration	isomorphic, homomorphic
Kind of Ralation between the Elements	linearly, non-linear
Degree of Determinability	deterministically, randomly
Behavior	statically, dynamically
Kind of Reference to Reality	real, ideal

Figure 3: Model Classification with Characteristics

In the context of computer-integrated manufacturing processes there are a number of rigid explanatory models. On the other hand there are also modelling languages that enable users to create individual models. Often, a skeleton model is equipped with a language that fills the skeleton model with tangible contents[Spu94].

3. METHOD AND MODELING LANGUAGE

The creation of non-trivial models requires suitable modelling methods. A modelling method consists of language constructs and a procedure describing how to apply the constructs effectively when creating models. A modelling method for quality-oriented design of business processes is presented in chapter 3 with focusing on modelling languages and in chapter 4 focusing on procedures.

A modelling language only includes language constructs consisting of language elements, e.g., letters, words and punctuation marks as well as formalized rules for the connection of elements, e.g., grammatical rules.

Method is the more comprehensive term. For the application of language constructs the modelling method provides a procedure going beyond the formalized connection rules of the language. These rules describe how to use the language – in the case of varying requirements – sensibly to describe a problem. Illustrated with an example, this means that our language consists of language elements, e.g., words, etc. It includes formalized grammatical rules, such as 'a sentence must have a subject, a verb and an object' or pronunciation rules. Rules, e.g., writing a summary of contents or an essay, are procedures. Procedures represent the most important steps of the approach towards attaining a certain goal.

Models are necessary to describe complex systems regarding relevant problems in a simplified manner. The modeling method helps to phrase the problem. Apart from descriptive constructs the method includes procedures that were created especially for the specific problem. The model should help users to solve this problem. Not every method is suitable for every problem. Modeling methods only vitalize the model's frame that has been generally defined [Mer93b].

4. BUSINESS PROCESS

Business processes are chains of corporate activities and their network-like relationships directed at the corporate objects 'product', 'order' and 'resource'. The relationship between business processes is like the relationship between customers and suppliers. The relations surpass organizational and system-related boundaries. The result is defined [Joc96]. A business process is a set of activities designed to yield a result that is of use to the company's customers. A business process is not restricted to a single function[HAM93, DS90, HC92, TJ92].

The essential descriptive features of a business process include the result, the input necessary to yield the result and the used resources. The structure of the company, created by the relations of the organizational units, is seen as influencing the efficiency of the business processes [Fah95].

Scheer understands business processes as a logical process of activities designed to fulfill a corporate task. The process involves the transformation of materials and/or information. Generally, a business process consists of several functions that are related and executed logically. Alternative and parallel activities are possible, too. To perform a business process the company needs staff members, suitable organizational structures, data, resources and information systems [All95].

5. QUALITY MANAGEMENT

The term 'quality' is multi-faceted. It has neither been pinpointed exactly, nor has it been defined by any generally acknowledged academic opinion.

During national and international standardization efforts, quality has been defined as 'the entirety of features of a unit in terms of its adequacy to meet predetermined and provided requirements' [DIN 8420].

The formula 'quality = technology + attitude', however, explains the development of quality: It is developed with the help of technical skills on the basis of the appropriate attitude [Kam90]. This may include a company-wide observation of quality. The consequent pursuance of this definition results in a quality description in the sense of Total Quality Management (TQM).

Quality management includes all managerial attitudes, intentions, objectives and measures that concern the achievement and improvement of quality. Multiple factors need to be considered, especially economical, environmental and legislative aspects. In addition, the companies should regard the wishes and requirements of their customers [Spu89].

The scope and substance of quality management are set out in a quality management manual (based on ISO 9000 ff.). In the course of a quality management system the measures are converted and applied.

6. BUSINESS PROCESS MODEL

A business process model is understood as the formal and semi-formal description of those features of a real or planned business process that are relevant to the purpose of consideration [Joc96].

The different aspects of a business process (functions, resources, organization, quality, costs and time) can be represented as different views onto one integral model. For this purpose, modeling methods are used. One example of such a modeling method, the method of Integrated Enterprise Modeling, IEM, is presented in chapter 3.

7. BUSINESS PROCESS DESIGN

The traditional way of corporate management to fulfill market requirements often consists of optimizing separated functions based on the existing organizational structure. This may lead to a number of problems with interfaces as well as to suboptimal results. The cross-functional study of continuous business processes if often neglected. Various views, such as quality, organization, information systems and costs (controlling), onto the company and its processes are neither integrated nor studied nor designed [Joc96].

'A model oriented towards the available or traditional organizational structure of a company cannot yield solutions that question just this organization' [War92].

Companies are required to realign the entire corporate planning and design along the value adding chain to accelerate the business processes. All relevant user views and affected departments must be considered and must be studied regarding their interaction.

Instead of emphasizing the organizational structure of a company, business process design attempts to plan and optimize the output of a company according to the requirements of the market. This is done integrally and oriented towards results and is detached from the organizational structure [BT92, Ham90, HT93, HAM93, DS90, Der94]. Business process design views the output of a company as a number of business processes that are interrelated logically and time-wise. The result of a business process is transmitted to one or more subordinate business processes. Here, the result is further processed. The entirety of business processes thus produces the output of the company that is relevant to the market [Wie92, EK93, NN94]. Business processes therefore 'communicate' with units of the corporate environment, e.g., customers or suppliers. 'Communication' in this sense implies the transfer of the results of the processes or the transfer of the input that is necessary to yield results. The results and the input may either consist of materials or information.

Business Process Reengineering, BPR, signifies the fundamental redesign of the business processes of a company. The goal of BPR is to achieve significant improvements in certain fields, e.g., processing times, costs or resource absorption [Ham94].

In contrast to the fundamental redesign achieved through BPR, continuous process improvement attempts to adapt and improve existing processes. The expenses for this method are not high. Therefore, appropriate measures may and must be performed more often. Relevant modifications may either imply step-by-step improvements regarding predetermined goals (e.g., reduction of costs) or the fast adjustment to a modified regulatory framework (e.g., changed customers' requirements) [All95].

During business process design business processes and their interactions should be studied and optimized. An essential part of optimization consists of reducing the complexity [Fah95].

Chapter 2

Requirements of Business Process Design

1. INTRODUCTION

The goal of business process design is to improve the customer orientation – especially by way of reducing processing times and improving the output regarding quality, delivery and costs. The new design of business processes may result in utilizing the integrative potentials of information processing technology, in integrating quality management, in improving the efficiency of the organization, and in incorporating indicators to illustrate cost structures. Business process should be oriented towards the corporate goals and the market requirements. The complexity should be reduced drastically. This will make business processes to a central element. The design of the organizational structure, the quality management system, the cost structures, the controlling system and the information systems should then be aligned with the business processes.

2. REQUIREMENTS OF ORGANIZATIONAL DEVELOPMENT

On the one hand, 'organization' describes the activity of organizing, on the other hand it is the structure that is the result of the activity. Organization regulates the distribution of tasks between people in goal-directed activity systems. The task of corporate organization is to determine the cooperative relations between the organizational units to achieve goals and to execute the fulfill necessary tasks.

The theory of organization distinguishes between the process organization and the organizational structure as two views onto an organization [Kos62]. The design of the organization is based on the production process and its sub-tasks. The process organization is being described through business processes. The tasks that are fulfilled in the production process are assigned to organizational units through the organizational structure [Kos62].

The organizational structure describes organizational units as
- restricted task areas by way of summarizing sub-tasks of a company and
- areas of decision and responsibility in connection with the assignment to productive resources.

The organizational structure determines the corporate division of labor and regulates the coordination of tasks and the areas of responsibility of the various organizational units. In a sphere of duty certain tasks are summarized in such a way that they can sensibly be assigned to one task station.

The creation of areas of decision and responsibility is necessary because manufacturing companies are goal-directed activity systems that have been designed by humans and that are subject to internal and external dynamics. Goals are set by people based on certain criteria. They are pursued by measures, i.e., by way of executing functions with resources. The system is dynamic because the corporate environment changes and because people with individual goals and methods try simultaneously to solve problems.

'Decisions' implies the determination of sub-goals and measures to achieve predetermined goals by way of selecting one among several alternatives. 'Responsibility' consists of assigning decisions and the resulting effects to people. Areas of decision emerge if areas of activities are equipped with authority. The areas of activity and the fields of decision and responsibility are assigned to specific people or teams. The smallest organizational unit is called a post. We can distinguish between executive posts, managerial locations (authorities) and staff posts. Posts are aggregated to departments and other organizational units of a higher order [Süs91].

The design of a business process is being determined by the entire task and the combination of the productive factors that are to fulfill the task. The key to design is to break down the total task of the production process into sub-tasks and sub-functions. This enables the designer to study the productive factors and their combination in detail.

A further requirement is to describe the information that is necessary to perform the business process as well as the representative data that needs to be exchanged between different posts within the company.

The multiple break-down of the business process into individual sub-processes allows you to identify basic functions and cross-sectional functions. Basic functions describe the production process itself. They include the geometric and technological design of the product, technical planning and provision of input factors, planning amounts and schedules and the production and assembly of products. Cross-sectional functions optimize basic functions according to certain criteria. For example, the cross-sectional function 'quality assurance' ensures that all functions are executed in such a way that the required quality is achieved [Süs91]. This concerns the production, production planning, resource planning, design and other basic functions. Other examples of cross-sectional functions include controlling, purchasing, personnel matters and data processing services. Various studies [Sel88, KCIM87, Sche87, Sche92] have described and analyzed the basic and cross-sectional functions of manufacturing companies.

The individual organization is designed by way of analyzing and synthesizing tasks, the organizational structure and the process organization [Kos62]. Gaitanides understands process organization as the design of the organization in which posts and departments are arranged in consideration of the specific requirements of the corporate processes [Gai83].

For industrial companies those organizational structures have proved to be sensible that are characterized by decentralized hierarchies of authority and thus by short information and decision processes. Successful companies are usually characterized by a simple organizational structure [Sch90].

To support the development of an organization efficiently business process design should support the analysis of tasks and the synthesis of organizational structures. For the synthesis of the organization the organizational structure, the process organization and decision and information structures should be optimized simultaneously.

3. REQUIREMENTS OF QUALITY MANAGEMENT

Labor-division has effected quality management badly. The operators of individual tasks lost their sense of responsibility for the quality of the total product. The responsibility for the quality of the total product was assigned to one specific area of the company, quality assurance. This meant that the responsibility was removed from those areas that really affected the quality [Lüb93].

The realization that most errors made in the course of the manufacturing process are caused in pre-manufacturing stages led to an expansion of preventive quality management measures (Statistical Process Control SPC, Failure Mode and Effects Analysis FMEA, Quality Function Deployment QFD etc.).

The central problem, the functional division of labor, however, was not solved through these measures. Therefore, the traditional and one-dimensional term 'quality' should be replaced by the comprehensive term 'quality management'. This term contains apart from the quality of the product the quality of internal activities that are not directly related to the product. This means that companies increasingly attempt to study and optimize corporate structures [Pfe93, TöM93].

It is, therefore, important to develop the readiness to constantly renew and improve the organization and the processes. This would be the prerequisite for quality design. In consequence, this means that quality management must be understood as a task concerning the entire company. The responsibility for quality should thus be retransferred to the individual staff members. This philosophy is also called 'Total Quality Management' (TQM).

For TQM, it is necessary to break up the traditional structures based on the division of labor. Tasks relevant to quality and organizational elements of customer-specific requirements should be designed and documented. Tasks cannot anymore be oriented towards departments and divisions, but should be oriented towards customers. Here, the internal and external customer principle should be applied. During staff orientation the responsibility for the quality of processes should be placed in the hands of the person that handles the particular process. Companies should rely on the knowhow and the active integration of staff members. This is the place for immediate improvements. However, independent acting requires clear orders and clear spheres of responsibility. In addition, the corporate structures must be comprehensible so that the individual staff member can find his role within the company. He could then realize the effects of his activities on the quality of the entire process [AWK93].

Process orientation – in connection with TQM – implies finding potentials for improvement in the processes and utilizing these potentials purposefully. Process quality is immensely important because only flawless processes result in flawless products and services. To achieve this goal, processes have to be known and transparent. Their interrelations should be obvious and have to be considered in any optimization measure. In addition, a continuous circular process of continuous improvements of the TQM principles must support any business process design [Kat94].

4. REQUIREMENTS OF INFORMATION SYSTEMS PLANNING

The information necessary for business processes should be represented by data. The most important data of the production process concerning its products, orders, resources and other variables should be represented during integrated enterprise modeling. Information technological systems enable users to
- process data,
- save data,
- transfer data and
- input and output data.

According to this arrangement, the systems may be demarcated from each other: The development of new information technologies can usually be assigned to one of the above-mentioned types. The other types are hardly affected at all [Fla86]. Information technological systems consist of hardware, i.e., physically existing devices and facilities, system software, i.e., programs that are necessary to operate the hardware and are not affected by the particular application, and application software, i.e., programs that execute data processing functions to support specific tasks.

Nowadays, when planning and introducing information systems that are designed to support business processes, functional and system technological aspects are in the focus of attention. However, it would be more desirable to plan and introduce systems that support or correspond to company-specific processes. This would lead to internal systems that could easily be adjusted to the specific processes – instead of adjusting the corporate processes to the selected systems [Süs91].

For this purpose, the supporting business processes and the required information should be described with the same method that is used to describe the information systems with their functional performance. A comparison of the business process model with the model of the information systems allows users to easily, quickly and precisely find out which of the modeled systems is best suited to fulfill the company-specific requirements of business processes. At the same time, the comparison identifies the modeled systems' need to adjust in those areas in which the systems show deficiencies.

To support flexible business processes, companies need information systems that can be adjusted to modified processes easily and flexibly. Corporate information systems are traditionally based on fixed processes that have either been predetermined by the system or that had been configured when initially installing the system. Re-configuring the systems is usually costly thus eliminating the possibilities for continuous adjustments [All95].

If a company is successful in linking modeling tools and operative information systems, it will not only be able to configure the information systems in such a way that they are oriented towards processes, but also to transfer changes of the process model directly to the information system and its implementation.

5. REQUIREMENTS OF CONTROLLING AND COST ACCOUNTING

Along with the quality, the costs are one of the most remarkable target of business process design. The increasingly accelerating technological development has led to changes in the corporate production and the creation of added value. This is because preparatory, planning, controlling and supervisory activities are increasingly contributing to the added value in all corporate areas. This results in a shift of the cost structures. In this regard, traditional cost accounting systems are insufficient. The incorporation of non-monetary cost determinants into consideration requires an expansion of traditional cost accounting to integral controlling, i.e., doing away with departmental sub-optima and attempting a corporate optimum [Loh91].

Indicators were already developed for controlling systems. Often, these developments did not take process design tasks into account. These activities should, however, be interlocked. This would ensure that users could recognize and describe present interdependencies much easier. The index number systems should not only be designed for the use of a central controlling department; all participants in the process should have access to these systems. They could then optimize decentralized processes easier and faster [All95].

Controlling is then understood as a sub-function of the management process. Its task is to collect, process and provide detailed information about the behavior and the results of past design measures. On the one hand, this is to ensure that decisions are indeed carried out, on the other hand, this supports and helps designers with a cost-related evaluation of different design measures and alternatives.

The introduction of distinguished indicators, that effect the costs to be pushed up, as relevant design features enables process cost accountants, on the basis of a detailed activity analysis and a description of the processes, to assign costs according to the true causative factors. By way of comparing actual and target indicators you may then trace deviations and inefficiencies back to the individual causal sub-process [Sci93, Sce93, Scu92, War93].

6. SUMMARY AND NEED FOR ACTION

To do justice to the multitude of relevant information and descriptive requirements of affected areas and to the views onto the company that were presented in chapters before, users should be enabled in business process design projects to adopt different views onto the company. Additionally, they would have to be put in a position to analyze and optimize the interactions and interdependencies [Mer95c]. Business processes and the relevant information should be described in one integral model. Information systems, the organizational structure, quality requirements and qualification requirements represent user views that are related to the core of the model. Process-organizational alternatives and changes may then be evaluated with regard to the effects on costs, quality, system support, organizational structure and the personnel-related qualification profile. The next two chapters contain descriptions of a modeling language and modeling rules that enable users to create such an integrated model for the design of business processes [Joc95a].

Chapter 3

Quality-Oriented Design of Business Processes

1. INTRODUCTION

To fulfill and to apply the requirements of quality-oriented business process design described in chapter 2 the business processes and the quality aspects are to be described with a modeling language. In this chapter, we describe the standardized modeling language of Integrated Enterprise Modeling (IEM) that supports an object-oriented, comprehensible description of business processes as basis of planning and conducting QM activities. This results in an increase of the quality of the corporate planning process.

The modeling language enables users to describe business processes and integrated quality aspects uniformly – both semantically and formally. As a result, all aspects are represented in one enterprise model. For representation purposes, we applied an object-oriented extension of EXPRESS (figure 4, also cf. chapter 3.2.4).

The modeling language and the descriptive requirements to integrate aspects of quality management are presented in chapters 3.2 and 3.3.

The modeling language is used to create business process models with an integrated quality management. To create such models faster and more efficiently, chapter 3.4 includes a description of pre-defined reference models for specific fields of application. These reference models are part of a structured model library. The development and application of the library are also illustrated in chapter 3.4.

Chapter 6 and 7 include practical examples of how to utilize the modeling language, the reference models and the model libraries.

2. MODELING LANGUAGE TO DESCRIBE BUSINESS PROCESSES AND QUALITY ASPECTS

2.1 Basic Constructs

2.1.1 Generic Object Classes

An object class is a pattern to describe objects, real things or occurrences, in a model. At the same time, the behavior of the objects is described through functions. The functions have exclusively been defined for the respective object class. Functions modify attribute values of objects (authorities) of the respective class. To define an object class, users also need to define functions that describe the behavior of objects of this class throughout the life cycle contained in the model.

All objects of one class are described by the same, common features (attributes). Modifications of objects are described by the same functions. When instanciating objects, specific values are assigned to the attributes of the objects. The condition (status) of an object is being described by the entirety of the values of its attributes. The execution of functions changes the status of the objects.

From an object class you may develop subclasses. The subclasses inherit all attributes of this superior class (parent class) – the objects of the subclasses are described by the same attributes. Additionally, each subclass is described in detail by class-specific attributes. The subclass is a specification of the parent class. For the subclass, users can newly define or eliminate individual attributes of the parent class. These changes are then passed down to all following subclasses.

IEM object classes are the basis of a comprehensive enterprise model. The unit of descriptive attributes and the defined functions to illustrate changes and interactions of the objects in the system 'company' ensure consistency and the reusability of the model. In a model, a company is described by objects, its relations and its behavior [Mer95d].

According to the purpose of the objects in the company we can distinguish the generic object classes 'product', 'order' and 'resource'. Modeling then takes place exclusively through objects of these generic classes or newly installed subclasses. Any aspect may then be illustrated as a view onto the objects of these classes. The sub-models that are on this basis defined as views are integrated through the relations to the same objects.

Quality-Oriented Design of Business Processes

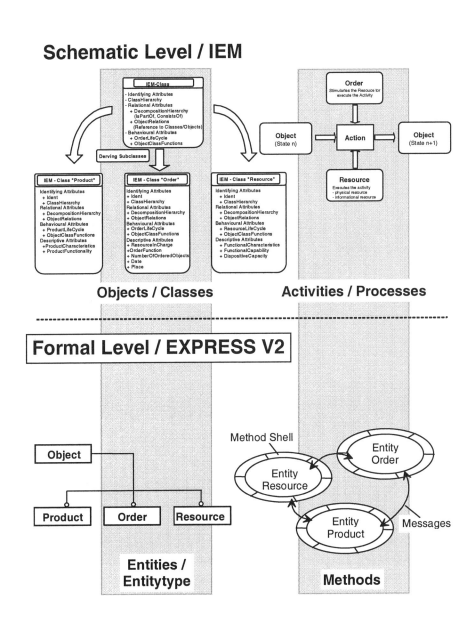

Figure 4: Integrated, Formalized Description of the Modeling Language

2.1.1.1 Class Product.

The goal of a company is the production and sale of goods. The objects of the class 'product' represent all intermediate stages of the product life cycle in the company – in terms of materials and in terms of information.

Step-by-step specification and specialization of the class 'product' allow users to define customary as well as company-specific subclasses of the class 'product'. Intricate product structures are described as 'is part of' and as 'consists of' relations between different subclasses of 'product'.

The objects of the class 'product' represent the products of the company. All necessary information required to manufacture the product, the information of the required characteristics and the quality parameters are represented.

Depending on the level of detail the description includes an illustration of all relevant states of the products, of the respective functions to process the products, of the logical sequence of the functions, and of all relevant relations with other objects and object classes.

2.1.1.2 Class Order

The object class 'order', along with their subclasses, serve to classify all kinds of corporate authorization efforts. The objects of the class 'order' and their subclasses are a summary of the information required to plan and control corporate functions, i.e., what is being done when and with which objects, in whose responsibility and with which resources. The interdependencies of the various kinds of authorization in a company are created through relations between subclasses of 'order'.

According to the different planning and control levels and according to the essential features of these levels, users may specialize the general features of the class 'order' to develop company-specific, hierarchically structured subclasses of the class 'order'. A possible classification consists in a division into global orders, i.e., customer and factory orders, orders that are restricted to certain areas, i.e., shopfloor orders and manufacturing orders, and orders that are restricted to certain activities, i.e., supply and procurement orders. These subclasses should be specified company-specifically in a class and/or a 'is part of'/'consists of' hierarchies. Possibly one might want to add further company-specific order classes.

Furthermore, users should describe the planning and control functions (methods) for the respective order classes, i.e., the processing of the above-mentioned information (processing of order objects) and the generation of further planning and control information (generating new order objects).

The objects of the class order represent the information in the company relevant to planning, controlling and supervising the processes.

These information concern the planning, authorization and control

- of functions to manufacture products,
- of resources necessary to execute the functions, including the necessary functions to prepare and supply the resources and
- of functions to process objects of the class 'order' itself.

The necessary information includes scheduling and quantity specifications, possible also locational data.

For each order class, the description includes an illustration of all possible states of the orders, of the respective functions to process the orders, and of the logical sequence of the functions.

The description should also include all those relations to other objects and object classes that are relevant to planning, authorization and control activities.

2.1.1.3 Class Resource

The object class 'resource' and its subclasses serve to classify and describe all aids – materials and information – that are necessary to execute or support corporate functions, i.e., machinery and equipment, organizational structures, data processing equipment and all kinds of documents. This list already contains possible subclasses of 'resource'. These subclasses should be further specialized and specified – according to the company-specific requirements.

The class definitions also contain all functions that are executed to achieve and maintain the efficiency.

The objects of the class 'resource' represent all things, facilities, persons and information that are able or necessary to execute functions.

The representation includes geometric, materials- and information-related features necessary to describe the ability to perform services or to enable or support the performance of services.

For planning purposes, users should also describe the ability to perform actions or functions (in the sense of the generic activity model) as well as the availability to perform services. The last aspect should include initial conditions.

For each resource class, the description includes the relevant states of the respective resources – according to the ability and availability to perform services –, the corresponding functions that are required to achieve or maintain the ability and availability (e.g., preparation, provision and maintenance) and the logical sequence of these functions. The description should also include all those relations to other objects and object classes that are relevant to the ability to perform services.

2.1.2 Generic Activity Model

To illustrate the functions of a company explicitly IEM has defined the generic activity model (Figure 5). The production of goods and the pertinent activities can be described according to a generic, object-related pattern: An activity is the purposeful modification of objects. Purposeful activities are based on indirect or direct planning and control efforts. They are performed by able contributors. These findings are the basis of their definition of the following functional constructs:
- The action is an object-independent description of an activity, a verbal description of a task, a process or a procedure;
- the function describes the transformation of objects of a class from one defined status into another defined status by way of performing an action;
- the activity specifies the controlling order for the status transformation of objects of one class that was described by a function. It also specifies the resources that are necessary to execute this transformation in the company – represented by an object status description.

The thus defined activity is a complete static description of functions at objects of a class and the relevant relations to other objects – the change of the objects is only described through the introductory and concluding states of the objects. The manner of the change is not being described.

Figure 5 illustrates the generic activity model and the three descriptive levels. The function on the object is illustrated by the introductory and concluding states of the objects and by the (object-neutral) action, connected by arrows from the left to the right. The objects that are to be modified are objects of the generic classes 'product', 'order' and 'resource' or of one of the subclasses. However, the objects are always only objects of a class whose common parent class belongs to one of the generic classes.

The authorization and control of the execution of a function by an object of an order class are illustrated by a vertical arrow from the top. This illustration describes the logical assignment (responsibility) of orders of one class for the execution of functions on objects to be modified.

Accordingly, the assignment of the resource responsible for the execution of the function on the object is illustrated by a vertical arrow from below. This also illustrates the logical connection between a function on objects of a class and the resource executing this function (in the real system). Time-related aspects and changes of the resource itself are illustrated by specific functions that modify or process these resources.

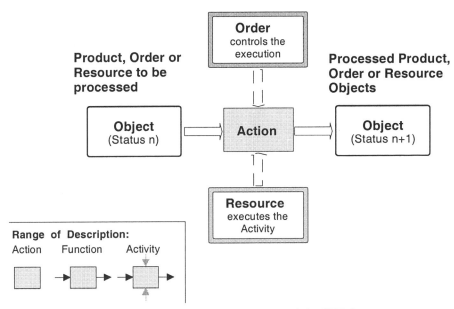

Figure 5: Generic Activity Model of IEM

2.2 Views

Models are descriptions of essential and relevant parts of the considered range of tasks. They do not duplicate reality, but represent a limited alignment to the considered sub-aggregate of reality. The suitable level of detail of a model is being determined by the intended use – the purpose of the model. A complete description of the model should contain statements of purpose, restriction and assumption (for modeling). This shows that the purpose of a model has to be in a predetermined relation with the purpose of the considered range of tasks.

2.2.1 Aspects and Views

While describing the observed aspects in the model, the modeler 'creates' a certain view onto the model. This view is determined by the modeling purpose and the modeler's point of view of the range of tasks. In other words, the description of one or more aspects corresponds with a certain view onto the object that is being studied. The modeler also has a general notion of how to structure and display the model of the observed facts.

Users in companies see the range of tasks and the model similarly. Their perception is determined by the necessity to perform or support a task. They use the model as a means to understand their scope of duties, i.e., to fulfill the requirements of their tasks.

2.2.2 Fundamental Views

Taking the (above-mentioned) purpose of a model and the close relationship between aspects (perception) and views (observation, discussion) as a basis it is obvious that viewers, modelers and users must have a common understanding of the purpose, the limits and restrictions and the intended usage of the model. Furthermore, they should share the common understanding of the essence of the task and of its representation in the model.

In each case, two aspects are of fundamental importance to understand the situation: the elements and the behavior of the system (scope of duties) and the view onto data and other information as well as onto the behavior of the company, i.e. an information-oriented and a process-oriented (functional) view, regardless of the applied modeling method.

Integrated Enterprise Modeling (IEM) is based on these fundamental views. The corresponding models of these views have been termed the view 'information model' and the view 'business process model' (Figure 6).

The view 'business process model' represents corporate processes and their interaction. It describes the possible states of objects, transitory states (functions) and the logical connection between objects depending on the transitory status (activity).

The object status descriptions, that are described by an activity, represent a certain amount of authorities (instances) of the corresponding class in exactly this described status. By way of specializing object classes in a certain application, users can include further detailed information. The specialization of order classes, for instance, allows users to describe control algorithms through methods of these classes and to represent the required data with suitable attributes.

The view 'information model' represents the object classes that are studied, the way these objects are structured and their attributes. It also represents those status descriptions of object classes that are relevant to the representation of processes in the view 'business process model'. This clearly illustrates the inseparable connection of the two fundamental views of Integrated Enterprise Modeling. The concept of object-oriented modeling allows users to define suitable classes and rules of inheritance. It thus ensures the openness and the expandability of a model as a basis of any number and kind of applications (cf. [Spu93b]).

Quality-Oriented Design of Business Processes

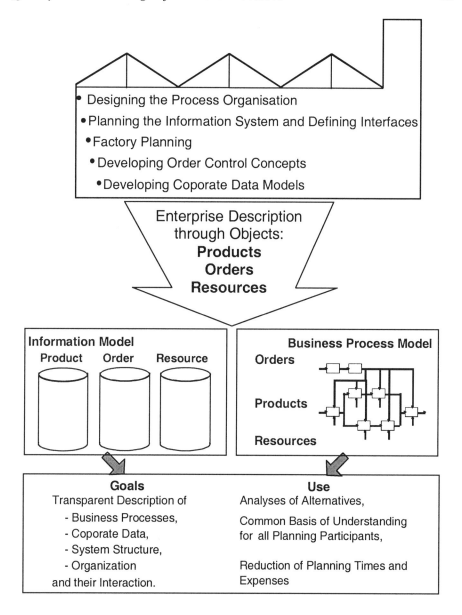

Figure 6: Main Views of Integrated Enterprise Modeling

2.3 Process Modeling

The basic concept of process modeling is the hierarchical structuring of the view business process model. For a certain field of the company the modeler only studies that information about the processes that is essential. Detailed studies are described on lower levels through connected subprocesses, global studies are described on superior levels through the total process, that is connected with other processes. The object states establish the connection between the levels and thus the consistency of this view. The object states describe the introductory and concluding requirements of the total process (Figure 7).

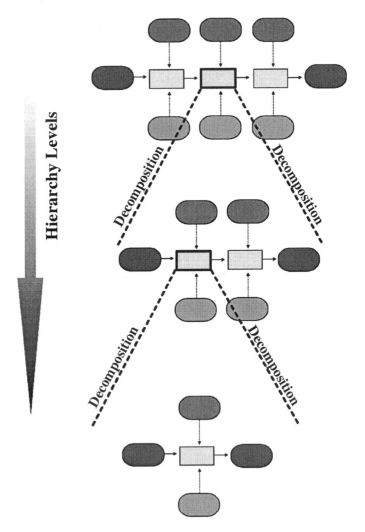

Figure 7: Hierarchical Structure of Business Process Models

Studied in its entirety, each corporate process can be described as an activity (cf. generic activity model, chapter 3.2.1). The activity is the

Quality-Oriented Design of Business Processes

underlying construct for the description of business processes. Activities (sub-processes) of one level of examination are connected through the object states of these activities: The ending status (the description of the objects that have been processes) of an activity is simultaneously the introductory status (the description of the objects that still need to be processes) of one or more other activities. The status descriptions of controlling orders and necessary resources represent the connections to corresponding order processing processes or to necessary resource processes. Depending on the respective task, users do not always have to describe all states of all activities in detail.

However, to represent the complex structures of business processes, modelers need further constructs. These are presented in the following.

We can distinguish the following types of connections (cf. Figure 8 through Figure 12):
1. a simple, sequential succession of activities,
2. a parallel connection of activities,
3. a case distinction as a selection among several alternatives,
4. a joinder to describe the union of various branches into one process.
5. Loops are represented as special forms of alternatives.

2.3.1 Sequential Connection

Figure 8: Sequential Connection

Connecting activities sequentially means that the activities are carried out successively.

2.3.2 Parallel Connection

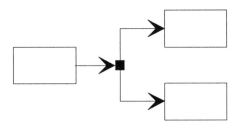

Figure 9: Parallel Connection

A parallel connection describes connected activities or processes that are carried out simultaneously, transferred or successively. The activity that follows two parallel functions can only be carried out of both prior functions have (been) ended.

2.3.3 Case Distinction

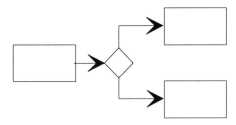

Figure 10: Case Distinction

A case distinction indicates that only one of two or more alternatives can be carried out – depending on certain conditions.

2.3.4 Loop

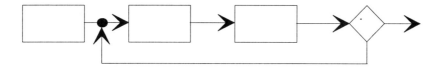

Figure 11: Loop

A loop (cycle) is a special case of case distinctions. Therefore, this type of connection is represented by the same graphic symbol. An element is repeated until the condition that has caused the cycle has become void.

2.3.5 Joinder

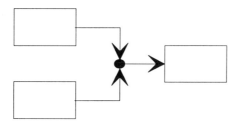

Figure 12: Joinder

A joinder designates the conclusion of the parallel or alternative execution of activities or processes. Chapter 7 and [Mer95f] contain an example for process modeling.

2.4 EXPRESS for the Formal Description of the Modeling Language

2.4.1 Introduction

The project STEP was initiated on an international level in the mid-eighties. STEP (Standard for the Exchange of Product Model Data) is the informal name of the standard ISO 10303 '*Industrial automation systems -*

Product Data Representation and Exchange' that was developed during this project.

Under the leadership of the ISO and in cooperation with numerous national standardization committees and industrial associations the goal was to develop an international standard that would consider all CAD/CAM data exchange aspects. An important goal of STEP was to standardize mechanisms to describe the product data of the entire life cycle of a product.

The chapter '*Description Resources*' of the ISO standard, part 11 '*The EXPRESS Language Reference Manual*' contains the formal language EXPRESS that was developed for this purpose (cf. List of Standards, Annex B). EXPRESS is a language to specify data and the structure of data. An important characteristic of EXPRESS is that it can be interpreted by a computer and that it can easily be read and understood by a person. The readability of data specifications and their development has even been enhanced by EXPRESS-G. EXPRESS-G enables the graphic description of sub-sets of EXPRESS, i.e., it is not possible to convert an entire specification from EXPRESS into EXPRESS-G. Even though EXPRESS resembles a programming language is has not been developed to describe executable programs. EXPRESS rather describes the data that is processed by a program.

Along with other numerous components, EXPRESS, edition 1, belongs to the initial release of STEP. Since 1994, it is one of the ISO standards (cf. List of Standards, Annex B). Edition 2 is currently being developed.

2.4.2 Specification of the Modeling Language in EXPRESS

To conform with international standardization efforts the data model of the IEM method was specified in EXPRESS [ISO10303-11]. This will facilitate a later description of the model information in standards such as STEP [ISO10303-1].

The IEM methodology allows users to define attributes freely. Figure 13 illustrates the appropriate EXPRESS pattern for classes (object_category), subclass relations (object_category_relationship) and attributes. Attributes are divided into the following three descriptive parts:

1. An initial part (attribute) containing the information that cannot be changed in the subclasses and instances (e.g., name and type of an attribute).
2. A descriptive part (object_category_related_attribut) that contains information that can be changed in the classes, not, however, in the instances, e.g., the value range and the default value of an attribute.

3. An assignment part (object_definition_related_category_related_attribut) that enables the assignment of a value to an attribute in one instance.

The three parts were defined as super type (ABS) to enable different features of attributes.

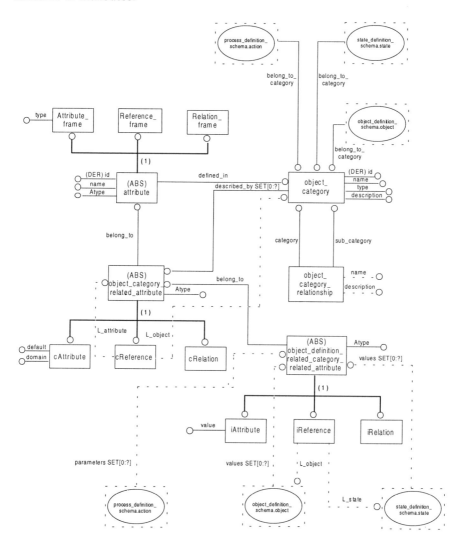

Figure 13: EXPRESS Model for Classes and Attributes of the IEM Method

The attributes are divided into three types of attributes. The 'relation' attributes should be seen as 'stand-ins' for the later description of relations. The 'reference' attributes refer to other attributes and objects contained in the model. 'Attribute' attributes are the third type. They describe data attributes that are defined by 'name', 'type', 'ID', 'domain', 'default' and 'value'.

The task of the entity 'object_category' is to define all IEM object classes that were developed from the three generic IEM object classes along with their attributes. Each IEM object class must contain a name and the name of the generic object class it belongs to ('product', 'order' or 'resource'). Therefore, the entity 'object_category' contains both entity attributes: 'name' and 'type'.

The entity attribute 'name' defines an object class; the entity attribute 'type' classifies the IEM object classes and assigns them to the three generic IEM object classes. It may occur that the name of an object class appears multiple times. It is not possible to use the same name to describe several object classes of the same generic IEM object class type. The pattern in the entity 'object_category' thus contains the entity attribute 'ID' that originates from the combination of the two above-mentioned entity attributes 'name' and 'type'.

The entity attribute 'ID' enables users to identify object classes of the same type clearly. To describe an object class with a short text you are provided with the entity attribute 'description' in the entity 'object_category'.

To represent the described attributes of an object class the entity 'object_category' contains the entity attribute 'described_by SET [0:?]'. It indicates that each object class may contain several attributes.

The objects of IEM represent objects of the real world. These objects are summarized in object classes. The 'Object_definition_schema' allows users to represent the explicit objects. As illustrated in Figure 14, it consists of the entity 'object' and refers to the external pattern 'Class_attribute_definition'.

The entity 'object' is described by several entity attributes. The entity attribute 'name' is not clear. It may appear several times in the model. The entity attribute 'type' allows users to define the type of an object. It has to originate from one of the three generic IEM object classes. Apart from a short description of the object with the entity attribute 'description' there is the identifier 'ID' to identify an object clearly. The identifier 'ID' consists of a combination of 'name' and 'type'.

Quality-Oriented Design of Business Processes

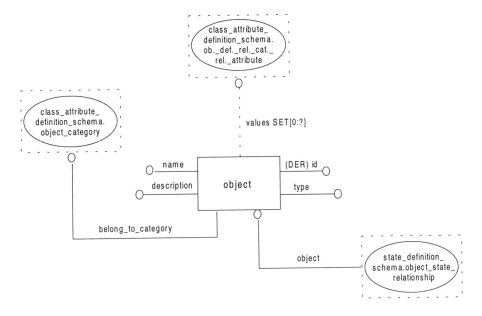

Figure 14: Object_definition_schema

The other entity attributes of this entity include 'belong_to_category' and 'values SET [0:?]'. The task of the first entity attribute is to establish a membership of the objects and the object classes. For this purpose, it refers to the external pattern 'Class_attribute_definition' in which the object classes are defined. The task of the entity attribute 'values' is to assign the object, that should be described, to its attribute values. The number of elements of this set corresponds to the number of object class attributes ('described_by SET[0:?]').

If an object class does not contain any attributes, the entity attribute 'values' has been defined as 'optional'. However, the possibility to identify the object class is till ensured by the entity attribute 'belong_to_category'. As illustrated in Figure 14, the entity attribute 'values' refers to the external pattern in which the attributes of an object are defined and the entity attribute 'belong_to_category' refers to the external pattern in which the object class of the object is defined.

After the description of objects and their connections to object classes and attributes, we now present status descriptions and their relations to object classes and attributes. The status descriptions are described just like objects and are represented through the entity 'state' as illustrated in Figure 15.

Each status description has a name that is defined through the entity attribute 'name'. The entity attribute 'description' allows users to describe the status description in a few words. The entity attribute 'type' has the same functions as in the entity 'object'. It enables users to assign a status description to one particular type of object class. The entity attribute 'ID' in the entity 'state' always consists of the attributes 'name' and 'type'.

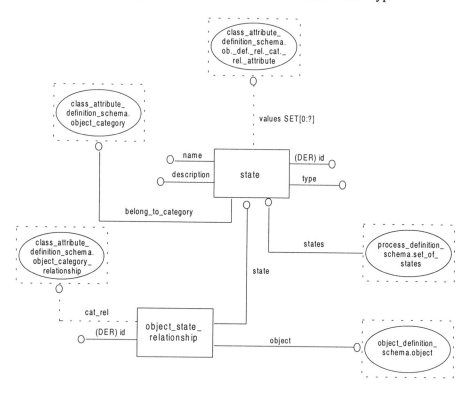

Figure 15: State_definition_schema

Further entity attributes in the entity 'state' are 'belong_to_category' and 'values SET[0:?]'. They indicate that a status description belongs to an object class. For this purpose, they both refer to the external pattern 'Class_attribute_definition' in which the object classes and their attributes as well as the attributes of the objects are defined.

Since object classes do not necessarily have attributes, the entity attribute 'values' has been defined as 'optional'.

The goal of the 'process_definition_schema' is to describe the logical connection of functions or activities of objects of the IEM object classes. At first, a description of the process elements (action, function and activity) is

Quality-Oriented Design of Business Processes

presented. Additionally, we describe the connection elements 'parallel' ('split'), case distinction ('decision') and connection ('join') that connect the process elements on the basis of certain conditions (Figure 16).

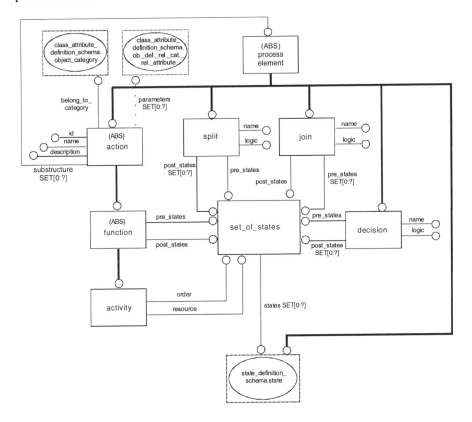

Figure16: Process_definition_schema.

The process elements of this schema are 'action', 'function' and 'activity'. An action describes a task in the model and contains defined data, e.g., the name and the ID. It refers to the object class in which it has been defined as well as to the features of the corresponding type of action. A function describes the change of an object of an object class in the model. For this purpose it refers to the object class and the respective action. Because the change of an object of an object class is described by way of indicating the status before and after the execution of a function a function also refers to status descriptions that are necessary to describe this particular change of the object. An activity serves to carry out a complete description of tasks in a model. For this purpose, it requires additional references to objects of the order and resource class when executing a function. These references refer to the status descriptions of these objects.

The connections between functions or activities are represented by the elements 'split', 'join' and 'decision'.

The entity 'split' should describe a parallel process between activities in the model. Apart from the entity attributes 'name' and 'logic' the entity contains the two entity attributes 'pre_states' and 'post_states'. They serve to describe the status descriptions before and after the connection. Before the connection there are numerous conditions: 'pre_states'. These status descriptions correspond to one edge before connection element. After the connection there may be several edges: 'post_states SET[0:?]'. These represent one set of states. The entity attribute 'logic' defines the logical function of the connection element.

The entity 'decision' should represent the case distinction in the model. It contains the entity attributes 'name', 'logic', 'pre_states' and 'post_states'. The entity attribute 'logic' defines logical operators that correspond to the logical function of the connection element, e.g., 'OR' and 'XOR'. The entity attributes 'pre_states' and 'post_states SET[0:?]' describe the status description before and after each connection.

The entity 'join' describes the combination of activities in the model. Apart from the entity attributes 'name' and „logic' it contains the two entity attributes 'pre_states SET[0:?]' and 'post_states'. They refer to the entity 'set_of_states' which contains descriptions of status descriptions.

3. INTEGRATION OF QUALITY MANAGEMENT

3.1 Process-Oriented Integration of Quality Management

The continuous adjustment of corporate structures and processes to rapidly changing market requirements is one of the major challenges of the nineties. The security and quality of processes must be guaranteed simultaneously. Certification according to the standard DIN EN ISO 9000 ff, which documents the ability for quality, are by now customary in many industries.

An enterprise model can be the basis for rationalization, the innovation of corporate business processes and certification activities. Enterprise models lead to the required transparency of value-adding and quality-relevant processes and the application of QM methods. Processes and QM can furthermore be documented on the basis of an enterprise model.

Process-oriented integration of QM thus means creating an integral enterprise model of all corporate business processes and integrating that information that is relevant to quality. The enterprise model thus contains the descriptions of QM, including the documentation of the twenty QM elements that are necessary for certification. Users may then develop the required QM manuals, procedure instructions and work instructions.

Traditional procedures only describe processes in an isolated manner – according to the QM elements. The advantage of this integral process study is the coherent view of all corporate processes. This may also be used for further improvement projects regarding times and costs.

When describing corporate quality management, we can distinguish methods of quality assurance (e.g., FMEA, QFD, SPC) and QM elements according to DIN EN ISO 9000 ff. that contain descriptions of the methodical application.

According to Pfeifer [Pfe93], QM elements can be divided into
– control elements that control and manage QM,
– process-accompanying elements that support the production of a product and
– process-related elements in which QM is being described within the production processes.

The integration of quality management into the enterprise model supports certification efforts through the creation of QM manuals, procedure instructions and work instructions. The QM manual contains company-specific descriptions of how to apply the demands of the twenty QM elements. The description of realized procedures for the QM elements requires the representation of:
– processes and detailed procedures,
– responsible organizational units,
– used and pertinent documents and
– resources of QM (e.g., testing devices).

At first, users have to get an overview of the corporate processes, especially the value-adding processes. The processes, the increase in value and particularly the connections between processes have to be specified step-by-step. The above-mentioned descriptive range is being completed iteratively. Connections of processes occur through instructions, through the procurement of resources, the same used documents, the same organizational units and through the same work systems.

To integrate typical QM procedures into an enterprise model you may use pre-defined models. The creation of QM, its documentation in the

enterprise model and the permanent improvement and evaluation of the QM are supported. The following aspects are represented:
- Process modules that describe QM procedures and support the development of an enterprise model,
- pre-defined models of control circuits and
- descriptions of indicators.

The following descriptive rules allow users to represent the above-mentioned information of QM in IEM models. Pre-defined modules for QM elements, QM methods, and control circuits for continuous improvement and the control of processes are also presented.

The illustrated procedures connect certification activities with corporate reengineering. We will show that the introduction of QM can be incorporated into an integral organization development process. Economic efficiency and certification can thus be achieved simultaneously; the management of change will be successful.

3.2 Descriptive Rules for Quality Management

Before attempting to introduce, maintain and improve a QM system you should first describe it. This chapter describes descriptive rules required to document a QM system.

To describe a QM system entirely it is sufficient to describe organizational units, documents, measures, procedures and processes.

3.2.1 Descriptive Rule 'Organizational Unit'

The organizational structure is understood as arranging the company hierarchically into so-called organizational units. These units vary in size, e.g., factory, central division, group and post [Wie86].

With IEM you can represent the structure of the organizational units, i.e., the structure of the organization in the information model and the relation between organizational unit and process in the process model.

Organizational units are object classes of the class 'resource'. Each company-specific organizational unit is defined as a resource class and is then arranged in a resource class tree (Figure 17) – according to its type (e.g., The department 'engineering' is arranged in the resource class tree as a subclass of the class 'department/division'.

To describe organizational units you need to define their features as attributes in the respective class. E.g., the attribute 'responsibility' is a feature of the resource class 'post'. The values of this attribute describe the status of the post in reference to the process, e.g., 'responsible' or 'in charge'.

Quality-Oriented Design of Business Processes

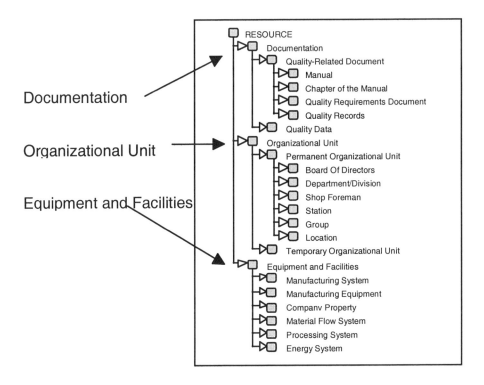

Figure 17: Representation of Organizational Units, Documentation and Resources in the Resource Class Tree

In the process model, organizational units are described as resource states, as, e.g., a Quality-agent. He is a post; therefore, he is defined as subclass of the class 'post' in the resource class tree. The process model contains a status of the class 'Quality-agent'. It is connected with the respective action through a connecting element (Figure 18). The values of the attributes describe the status unmistakably. For example, the value of the attribute 'responsibility' determines whether or not the 'Q-agent' is responsible or in charge of 'audit planning'.

3.2.2 Descriptive Rules 'Documentation'

This rule concerns the description of all quality-related documents and quality data that exist in the company. This includes the QM manual with all its chapters, QM procedure instructions, QM work instructions, standards, guidelines, regulations and quality notes.

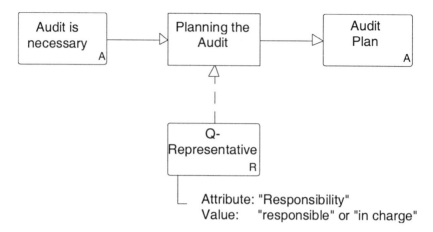

Figure 18: Description of an 'Organizational Unit' in the Process Model

The basic approach to representing the documentation never varies. At first, the user has to define a specific documentation as resource class. According to its structure, it is arranged in the resource class tree (Figure 17).

Each documentation is described through features. Such features include the version, the document number, the type, but also text-based descriptions of the purpose of the document. The features are defined as attributes in the classes. They can then be described with document-specific values.

The process model represents the relation between the documentation and the activity the documentation is needed for. Here, the documentation is represented as a resource status and is connected with the respective action through connecting elements. The procedure is analogous with the procedure applied with the organizational units (Figure 18).

3.2.3 Descriptive Rules 'Resources'

Resources include machines, facilities and all other devices needed in a work system to fulfill a task [REFA93]. From the viewpoint of QM, testing devices and labeling systems should also be regarded as resources. The procedure is analogous with the procedure applied to describe the documentation.

3.2.4 Descriptive Rules 'Procedures'

QM requires descriptions of procedures as a working basis for the middle management. A procedure is described by a sequence of activities (in IEM,

Quality-Oriented Design of Business Processes

these are also called 'functions') that are characterized by having measurable input and output. Therefore, we can also determine a change of status.

Each activity is described by a name., that is represented by the action name, and by an introductory and concluding status. The values of the features that describe the respective status are contained in the states themselves as attribute values. Figure 19 illustrates an example of this procedure.

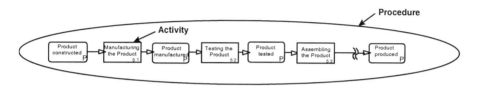

Figure 19: Description of the Procedure Instruction 'Produce' in the Enterprise Model

The connecting elements of the modeling language (cf. chapter 3.2) allow users to represent any sequence of activities.

3.2.5   Descriptive Rules 'Processes'

Haist/Fromm [Hai93] define the term 'process' as the interaction of people, machines, materials and procedures arranged to perform a service or to manufacture a certain product.

From this definition we can infer the following requirements:
1. Users have to be able to define the introductory and concluding state of a process. The concluding status is the result of the process that can be measured by features. The introductory status describes the initial status of the process.
2. People, machines and procedure have to be represented.

The relations between 1. and 2. should be comprehensible. IEM represents a process as shown in Figure 20. Each process can be divided into a sequence of tasks. IEM calls the interaction of tasks, orders and resources an activity. The term activity does not, however, imply a certain extent or importance of the results.

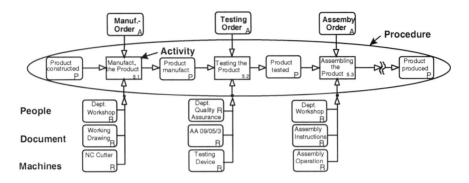

Figure 20: Representation of a Process with IEM

If the amount of data is reduced, the processes get clearer. IEM supports this through the possibility to create hierarchical levels. This is also called 'decomposition' (cf. chapter 3.2). The descriptions of processes would then only contain the name of the process, the introductory and concluding status, the orders and the resources (Figure 21).The process name is represented in the action box (rectangular box). According to the goal of the description, the introductory and concluding status of the process may either belong to the class 'product', or the class 'order' or the class 'resource'.

People, machines and documents are represented as resource states. Resources are necessary performers that are required to achieve a certain process result. According to the definition of a process, resources contain: people as part of the organization, machines as part of the equipment and documents, that are not listed explicitly.

The description of the order states represents the control of the processes. Arrows illustrate the connection between the components of a process graphically. Uninterrupted arrows illustrate which introductory status leads to which concluding status or result. Dashed arrows indicate whether the status is controlling (dashed arrow from the top) or supporting (dashed arrow from the bottom).

Quality-Oriented Design of Business Processes

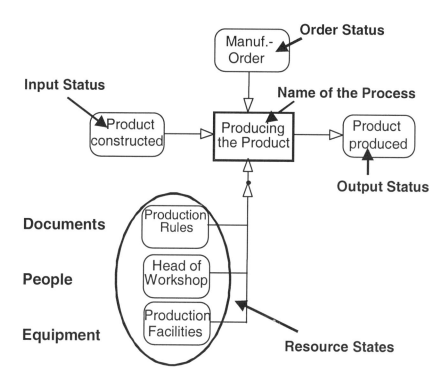

Figure 21: Description through Resource States

3.3 QM Elements as Process Modules

Central requirements of a comprehensive quality management include designing, representing and documenting the quality-related processes in the company. This is due to the fact that it is getting increasingly important for worldwide suppliers to prove their capability for quality. DIN EN ISO 9001 places the greatest demands on the QM system. It is considered *the* central standard in the field of quality assurance.

Pre-defined models of corporate quality management processes that are based on requirements of the standard DIN EN ISO 9001 can support the development of a corporate QM system within the value-adding chain. While serving as a model for process models, they should facilitate the modeling of QM processes. If there is, for example, a reference model for procurement processes a company can adjust this model to its requirements to create a company-specific process model for procurement. Reference models for quality management are called process modules that are incorporated into the process model of a company. The company-specific model can be used to document the QM system.

Corporate QM systems are effected by various internal and external aspects. There cannot be a universally suitable (standardized) QM system. DIN EN ISO 9001 thus does not contain any solutions, but demands that have to be fulfilled by a company-specific QM system. The demands indicate which object needs to be modified and what (activity) needs to be done (for example, defective parts have to be tested). These demands lead to tasks (e.g., testing defective parts) that a certified QM system must contain. By way of logically connecting the tasks of individual elements, users can describe processes and develop process-oriented models. These models describe standardized procedures and do not refer to a methodology to fulfill demands. The models do not contain specific statements concerning how, when or where to carry out activities and who the person responsible is. This company-specific information should be specified by the user when incorporating the QM process modules into the user model.

The demands of DIN EN ISO 9000 ff. concern all activities that effect the quality of a product or a service. They concern the entire corporate organization and effect each staff member. Therefore, the introduction, description and maintenance of quality management are complex tasks that require users to divide the QM system up into manageable subprocesses and to arrange these processes logically. To develop and use QM process modules on the basis of IEM it is thus necessary to
- restrict the process modules and to
- integrate the process modules into the corporate model.

The structure for the development of a QM system, predetermined in the standard DIN EN ISO 9001, divides the system into twenty elements. Each element examines a basic task of QM (e.g., contractual agreements, testing devices, etc.). Users can develop tasks from each element and describe them in an integral process model. The restriction of the QM process modules can, therefore, be carried out on the basis of the restriction of the twenty elements of the standard DIN EN ISO 9001.

The integration of the process modules into a flexibly applicable IEM enterprise model on the basis of the value-adding chain required a detailed analysis of the individual elements. The elements do not exclusively concern the production of goods in a company, but also describe processes that are not connected directly to the production (QM elements, 'training', 'testing devices', 'internal audits', etc.). Users must also consider the IEM classification of corporate objects (product, order and resource) when integrating process modules. For example, the QM element 'products provided by the customer' refers to the object class 'product' and the element 'testing devices' to the object class 'resources'. When integrating

Quality-Oriented Design of Business Processes 45

process modules into a flexibly usable enterprise model, users should, therefore, apply different rules.

Pfeifer [Pfe93] divides the twenty elements of DIN EN ISO 9001 into:
- process-related elements,
- process-accompanying elements and
- control elements.

Based on this division, the process modules are divided into process-related process modules, process-accompanying process modules and controlling process modules (Figure 22).

Process-related process modules refer to processes that can be directly assigned to the value-adding chain, from offer processing to maintenance (for example, 'examining contracts'). The goal is to integrate activities that serve to fulfill the requirements of the considered element into the suitable stage of the enterprise model.

Process-accompanying process modules refer to processes that have to be fulfilled at several points of the value-adding chain (for example, 'testing devices'). This means that these modules should be integrated into the hierarchy of the IEM model at a suitable position.

Controlling process modules refer to management tasks that effect the necessary conditions, that examine the effectiveness of quality-assuring measures and that steer the measures in the desired direction for example, element 1: 'management responsibilities'). The actions taken to fulfill the requirements of these elements modify the objects of the IEM class 'order' and thus correspond to the order view.

This determination facilitates the use of QM process modules according to DIN EN ISO 9001. The process-related process modules describe individual stages of order processing from the point of view of the QM system. The process-accompanying process modules are sub-assigned to the process-related elements. They refer to indirect production activities and possess a large number of interfaces with the process-accompanying process modules. They are only carried out if a pre-defined event occurs. This event can occur at different points of the production process. For example, the process module 'managing defective parts' is always applied if a defective product status occurs within the production process.

The controlling process modules create orders for other QM elements and are modeled within the order view. Predefined process modules for quality management effectively support users that attempt to develop and document a corporate QM system. However, a QM system is not a static system; it is a system that must be stabilized and improved constantly. The following chapter presents two methods that help users to improve the effectiveness of the QM system further.

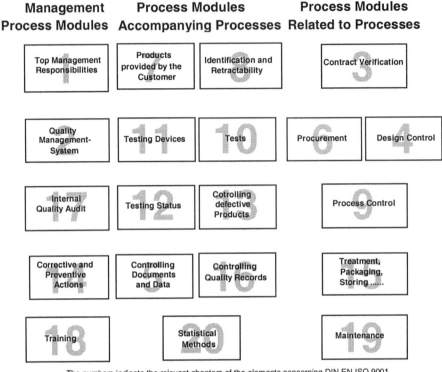

Figure 22: Delimitation of the Process Modules

3.4 Support of the Corporate QM System through Control Circuit Models and Process Models to Derive Indicators

3.4.1 Control Cycle Models

Nowadays, corporate QM does not any more focus exclusively on the unique design and documentation of the quality-relevant processes in the company. It rather concentrates on continuously adapting and developing all processes in a company. The typical tasks of quality management are
– controlling and assuring the quality-relevant activities and processes and
– continuously improving the quality-relevant processes.

Quality-Oriented Design of Business Processes 47

The essential goal of a company should be to manufacture top-quality products. To attain this goal an organization "has to ensure that those technological, organizational and human factors are controlled that effect the quality of its products, be it hardware, software, procedural products or services" [DIN94]. Controlled processes are the basis of continuous process improvements.

A process should be simple and secured by a control circuit if users wish to safeguard and command it. Pfeifer's comparison of the tasks of control system engineering with the tasks of quality assurance made clear that in principle the model of control system engineering can easily be transferred to quality assurance [Pfe93]. The goal of control is to keep certain initial values (controlled variables) of activities and processes on the level of predetermined target figure. For that purpose, the output figure that should be controlled is measured and then compared with a target figure. On the basis of the difference of the two figures and with certain rules (controllers) users develop short-term measures to secure the activity or the process.

The advantage of such a procedure is that errors are recognized as early as possible and that rework expenses are kept low. The control steers the individual process and reacts to trends or deviations from the scatter diagram of the initial value. Variance is here an essential indication of the total behavior of the process [Kir95]. Examples of the short-term control of a process with measures of QM include self-evaluation, statistical process control with quality cards or Design FMEA.

To model control circuits, users have to identify those functions/activities in the model that greatly effect the quality of the products (e.g., procuring components that concern the security). The control circuits must then be specified and integrated into the IEM model. Control circuit modeling can also be supported by predefined control circuit modules. The application of these modules corresponds to the application of QM process modules. Apart from the use of control circuit modules from control system engineering the modules can also be used to represent the applied procedures and methods of QM (self-evaluation, SPC, QFD, C-FMEA, etc.).

3.4.2   Process Models to Derive Indicators

Apart from controlling local activities or processes with process-internal control circuits the key function of corporate QM also includes identifying long-term weaknesses of the processes and planning and realizing improvement measures. On a long-term basis, quality requirements should be fulfilled purposefully. The general quality level should be optimized. This aspect is further specified in DIN EN ISO 9004, part 1: „If a QM system is realized management should ensure that the system facilitates and promotes continuous improvements of the quality" [DIN94].

Continuous improvement is a long-term process. It concerns inter-departmental cooperation organized like internal and external customer-supplier relations. This set-up requires process-overlapping control circuits. Because the measures that are to be taken cannot be developed directly from the control values and because decision processes have to be run through, users cannot determine unmistakable regulations. Therefore, the application of the above-mentioned control circuit modules is not possible.

To recognize weaknesses of the current situation and to introduce improvements, the necessary information – as decision support – must be available at the correct location and must be evaluated with suitable methods. For that purpose, users should collect the appropriate data from individual processes and condense them into indicators. To model this aspect, users have to develop process models to derive indicators. These illustrate the connection of control figures of different subprocesses that concern the corporate performance and document the process of continuous improvement.

To measure these control circuits, data collection activities are integrated into the model at predefined locations. Within an evaluation process the data is condensed into indicators that are the basis of the development of improvement measures. For frequently used evaluation processes, you can use predefined process modules (process FMEA, process capability studies, field data acquisition and processing, quality costs acquisition and processing and suppliers' evaluation etc.).

3.4.3 Example of the Application of Control Circuit Models and Indicators

While checking a procurement order, it is noticed that the ordered product has not been specified sufficiently. The person responsible requests the desired specification, modified the procurement order and sends it to the supplier. This short-term control of the initial variable can be represented in the process model with a control circuit model.

An example of the long-term improvement of a process is the reduction or the increase of the testing expense of subcontracted parts in the receiving department. To decide whether the reception procedures should be changed staff members need suitable indicators (e.g., number of defective parts in relation to the number of ordered parts) These indicators are determined through data that is collected in different processes (procurement process: number of ordered parts; process of incoming merchandise: number or defective parts).

4. REFERENCE MODELS AND MODEL LIBRARIES

4.1 Introduction

Reference Models enable users to create models more quickly and more efficiently. They provide predefined model structures that merely have to be adjusted and specified. In chapter 3.4.2 and chapter 3.4.3 we present two reference models for two applications (order processing and quality management).

The reference models are part of a model library. The library is structured according to the following criteria:
- range of application (order processing, quality management, ...),
- industry (mechanical engineering, electrical engineering, ...),
- manufacturing type (one-of-a-kind production, batch production, mass production) and
- type of organization (decentralized, segmented, ...).

The following aids are provided:
- Basic models of typical model structures, that can easily be adapted and that represent basic principles, e.g., of an application,
- examples of typical model structures that contain solutions for practical design problems, that do not, however, contain company-specific peculiarities, and
- modeling rules for the efficient creation of models utilizing the typical model structures of the library.

A reference model is structured according to the principle of modularity and consists of four components: object classes, basic functions, basic structures and process structures (Figure 23).

Reference models are based on the definition of object classes and the generally acknowledged description of functions that represent the treatment of defined classes. The classes and functions are used to describe basic structures of applications. To describe process structures, users then apply the classes, functions and basic structures. To develop specific models, users can use and, if possible, adapt, all four components.

4.2 Reference Model 'Order Throughput'

4.2.1 Elements of the Reference Model

The reference model 'order throughput' was developed to structure and optimize the order processing and order control processes with regard to processing times, cost structure, quality requirements, resource expenses and the support of information systems.

Reference models consist of basic models, example models and modeling rules. Basic models are predefined model structures that can be used to create complete models. The component 'basic model' is further detailed into generally accepted modules. All modules can be combined and adjusted to specific tasks. This facilitates the creation of specific models greatly [Mer94].

The reference model is created according to the principle of modularity (cf. chapter 3.4.1). Modules are:
- order, product and resource classes,
- basic functions and structures of order processing and
- business process structures.

Due to the integral study of business processes, product and resource classes are usually defined neutrally. With resources, the main focus is usually on organizational units and the assigned capacities. The structure of order classes is explained in detail in chapter 3.4.2.2.

On a general level, the basic functions of order processing represent typical functions of order processing. The basic structures of order processing consist of these basic functions. The structures either represent typical processes, such as order preparation and tracking, or typical organizational forms of order control (centralized or decentralized, push or pull principle).

The basic structures are used to model the order processing and control structure of the typical business process structure. The basic structures are oriented at the morphological features of order handling.

Quality-Oriented Design of Business Processes 51

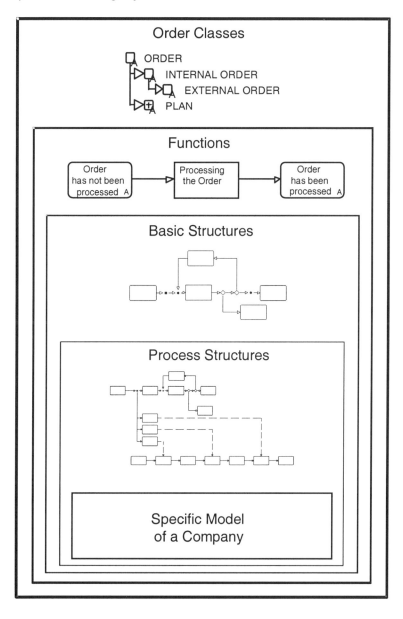

Figure 23: Principle of Modularity of a Reference Model – the Example of the Reference Model 'Order Throughput'

4.2.2 Classification of Orders

In this chapter, the orders are classified, the individual classes are defined and their attributes are described or explained.

4.2.2.1 Class 'Order'

The class 'order' represents the information that is common to all kinds of orders and that is necessary to plan, control and supervise corporate processes.

The class 'order' contains the following attributes:

Order Name. The attribute serves to name the order, e.g., design order, manufacturing order, etc.

Order Ident. The attribute serves to identify the order clearly by an identifier that is only awarded once. For example, you can represent order numbers.

Order Item. The attribute describes the performance requested by the orderer. An order may consist of several items. Each item is described by the following attributes.

Object. The attributes name the objects that have been ordered or refer to an individual object.

Function. The attribute refers to the function that is to be executed with the objects or to the function that is required to fulfill the order.

Target Amount. The attribute represents the parameter of the amount of objects ordered.

Actual State. The attribute represents the parameter of the objects that have been processed according to the order after the functions have been executed.

Unit of Quantity. The attribute indicates the units of quantity for targeted and actual amounts (e.g., piece, t, kg, lb., etc.).

Date_for_Item. The attribute indicates the expected finishing date for the order item. The date may be different for each item. The attribute is awarded by the orderer.

Supplier. The attribute refers to the resource (e.g., a person, an organizational unit, a system, etc.) that is responsible for executing the task specified in the order item. This resource is also responsible for announcing the execution of the task to the releasing order.

Processing Status. The attribute 'processing status' describes the processing and scheduling status of the order. Each newly executed function resets the value of the attribute. Each order runs through the following functions: preparing an order, breaking up an order, scheduling an order, arranging an order, recording and processing feedback, and – if necessary – processing disturbances. An order may also be suspended. Correspondingly, the value of range contains the values: prepared, broken up, scheduled, released, disturbed, canceled and finished.

Subordinate Orders. The attribute describes which follow-up, sub-, split or concentration orders were created – by way of breaking up or summarizing – to process an order.

Senior Order. The attribute indicates the orders this order has been developed from – through order breaks-ups or through summaries.

Feedback. To pursue orders and to manage the order processing, the execution of an order and its subordinate orders should constantly be controlled. With internal orders that control the execution of an action directly, the attribute describes feedback dates and results and indicates whether these have to be announced to the senior orders.

4.2.2.2 Class 'Internal Order'

The class 'internal order' is a subclass of the class 'order'. Internal orders are created if external orders (e.g., customers' orders) are broken up, if schedules are broken up (e.g., production schedule) or if senior internal orders are broken up. This means that internal orders are only created n the basis of existing (internal, external or schedules) orders.

In addition to the attributes of the class 'order' the following priority and scheduling attributes are defined:

> Order Priority. Generally, two parameters are decisive to determine the sequence of orders in the waiting line: the priority, i.e., the importance, and the interval to the latest beginning date of the order. The attribute 'order priority' indicates the priority of the order.

> Expected Final Date. The attribute indicates at what time the entire order is supposed to be finished. If there are several items with varying expected final dates the latest finishing date of an item is the latest final date for the entire order.

> Earliest Beginning Date. The attribute indicates the earliest beginning date for order processing. The scheduled date and the time at which the order processing actually starts cannot be earlier.

> Latest Beginning Date. The attribute indicates the latest beginning date for order processing. The scheduled date and the time at which order processing actually starts cannot be later. Otherwise, the final date cannot be kept.

> Earliest Finishing Date. The attribute indicates the earliest finishing date for order processing. The scheduled date and the time at which order processing actually ends cannot be earlier. If the earliest finishing date is later than the expected finishing date this fact must be reported to the subordinate order as a disturbance.

> Latest Finishing Date. The attribute indicates the latest finishing date for order processing. The scheduled date and the time at which order processing actually ends cannot be later.

Quality-Oriented Design of Business Processes 55

Scheduled Starting Date. The attribute indicates at what time order processing should start. This date is determined in the course of fine-tuned order planning.

Scheduled Ending Date. The attribute indicates at what time order processing should end. This date is determined in the course of fine-tuned order planning.

Actual Starting Date. The attributes indicate the time at which order processing actually started.

Actual Ending Date. The attribute indicates the time at which order processing actually ends.

4.2.2.3 Class 'External Order'

The class 'external order' is a subclass of the class 'order'. External orders go beyond the borders of the company; they represent orders and inquiries of customers and suppliers.

For external orders we have additionally defined the following attributes:

Customer. The attribute indicates the resource (customer or supplier) that releases the order. This resource has to be informed about order disturbances or the completion.

Delivery Address. The attributes indicate the delivery address for the objects that were ordered.

4.2.2.4 Class 'Schedule'

The class 'schedule' is a subclass of the class 'order'. The class 'schedule' represents information that is necessary to process schedules.

The class 'schedule' contains the following attributes:

Scheduling Resource. The attribute refers to the resource that is responsible for the creation of the schedule. This resource is responsible for recording and processing feedback of the production schedule.

Start of Schedule. The attributes indicate the time at which the schedule begins to be valid. At this date, old scheduling dates are either replaced by new ones or the schedule is valid for the first time.

End of Schedule. The attribute indicates the time until when the schedule can be considered the controlling order.

Rescheduling Date. The attribute indicates the time at which the schedule needs be rescheduled. This date is before the time indicated by the attribute 'end of schedule'.

4.2.3 Order Processing Functions

The basic functions of order processing represent – on a general level – typical functions of order processing. In this chapter, we present the functions order break-up and scheduling – as a substitute for the basic functions for order processing described in the reference model.

With the help of these predefined functions, users can easily create control circuits and complex order processing processes. If necessary, the functions that were defined for order reference classes have to be adjusted for specialized order classes or have to be expanded for specific requirements of individual enterprise models.

4.2.3.1 Order Break-Up

An order can be broken up into subordinate orders (Figure 24). The subordinate orders then assume the control of the real order execution or are further broken-up into subordinate orders. The rules for the creation of subordinate orders are company-specific.

The function decides whether, and if yes which subordinate orders have to be created. It then creates these orders. Subordinate orders can be created in three ways: through the creation of sub-orders, the creation of batches and the creation of follow-up orders. These three ways can be considered as separate order break-up functions.

Quality-Oriented Design of Business Processes 57

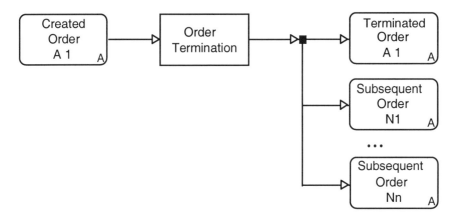

Figure 24: Function 'Order Break-Up'

4.2.3.2 Order Scheduling

The task of scheduling functions (Figure 25) is to schedule the following dates: expected finishing date, earliest starting date, latest starting date, earliest finishing date, latest finishing date, scheduled starting date and scheduled finishing date.

Schedules require values for the attributes 'start of schedule', 'end of schedule' and 'rescheduling date'. For subordinate orders, users also need to determine the value of the attribute 'feedback'.

If an order with a predetermined finishing date cannot be scheduled anymore the processing is interrupted. The occurrence is reported back to the senior order where the further procedure is decided upon, i.e., whether the order is canceled or whether new dates are set for this order.

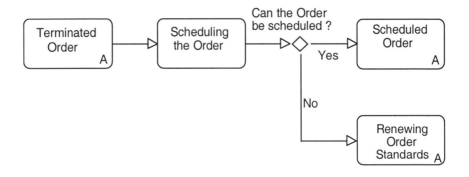

Figure 25: Function 'Order Scheduling'

4.2.4 Basic Order Processing Structures

The basic structures of order processing that represent typical processes, such as order tracking and preparing processes, on the one hand and typical organizational forms of order control (centralized or decentralized, push or pull principle) on the other hand, have been combined from the basic order processing structures. These basic structures can easily be adjusted for specific enterprise models. An example of a basic structure is contained in Figure 26. A description of all basic structures is contained in [Mer94].

4.2.4.1 Basic Order Processing Structure for Orders with Subsequent Orders (GS1)

This basic structure can be used for the development of an enterprise model for orders that – from the beginning – include subsequent orders, e.g., customer orders or production schedules.

The order may either be a schedule or an internal or external order. The processing status 'created' describes an order with an expected deadline and a list of order positions that must further be broken up.

The function 'order break-up and scheduling' determines the attributes 'subsequent orders' and 'senior orders'. In addition the order positions and the expected deadlines of the subsequent orders are determined.

4.2.4.2 Basic Order Processing Structure for Actions of Controlling Orders (GS2)

This basic structure can be used for the development of an enterprise model for orders that control an action and need not be broken up further.

Quality-Oriented Design of Business Processes 59

The considered order can either be an internal or an external order. The processing status 'created' described an order that has been broken up so far that it can be released without further break-ups. The order positions and the expected deadline of the order are predetermined.

4.2.4.3 Basic Structure of General Order Processing (GS3)

This basic structure can be used for the development of enterprise model for orders that either control an action or are broken up into further orders.

The Basic Structure of General Order Processing (GS3) is combined of the basic structures GS1 and GS2.

The function 'order break-up and scheduling' decides whether the order is further broken up (GS1) or whether the order can be scheduled in detail and then released (GS2). If an order is initially carried out but the release is disturbed, a continuation of the order enables you to break up the order this time.

4.2.5 Process Structures for Order Processing

This chapter describes – in a general form – basic process structures of order processing. These process structures are a combination of classifying orders, functions of order processing and basic order processing structures. The process structures can easily be adapted to specific enterprise models.

The general process structure describes the idealized production of a product – beginning with the design, including task scheduling, procuring and manufacturing activities and ending with the shipment of the finished product.

The basic structures that were described in the previous chapter are used to model the order processing and the order control structure of typical business process structures. These basic structures are oriented towards morphological order processing features. On this basis, we have defined three basic models for business process structures:
– customer-related production,
– anonymous pre-production – final production related to customer orders and
– stock production.

4.2.5.1 Make-to-Order Production with Individual Orders

In the general process structure for make-to-order production with individual orders all actions are controlled by orders that originated from the break-up of a customer or its subsequent orders.

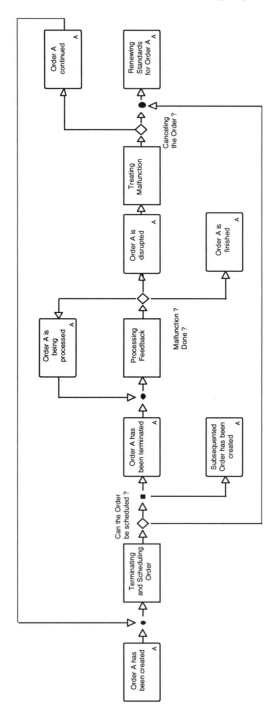

Figure 26: Example of a Basic Order Processing Structure

4.2.5.2 Stock Production

In the general process structure for stock production all actions are controlled by orders that originated from the break-up of the production schedule. Only shipping orders are customized.

4.2.5.3 Anonymous Pre-Production / Final Production related to Customer Orders

In the general process structure for Stock Production

In the general process structure for stock production all actions are controlled by orders that originated from the break-up of the production schedule. Only shipping orders are customized.

Anonymous Pre-Production / Final Production related to Customer Orders all actions – except shipping actions – are controlled by orders that originated from the break-up of the long- and short-term production schedule and its subsequent order. Afterwards, all actions can be controlled again by customized orders. The process structures are illustrated graphically in [Mer94].

In the following subchapter we describe how to apply and integrate a specific modeling method to reduce the modeling expenses for technical order processing considerably.

4.3 Model Library Quality Management

The model library quality management supports users attempting to create and design a corporate QM system. The library contains predefined class structures for the IEM objects 'product', 'order' and 'resource' (object class library) as well as classes of QM process modules (process library). The classes of QM process modules contain predefined process models that were developed from the requirements set by the individual elements of the standard DIN EN 9001 and that correspond to the conversion of QM elements. The QM process modules are imported into the enterprise model of the user through simple mechanisms. They can then be modified easily. For this purpose, classes or objects can be renamed, moved or deleted – according to the actual corporate processes.

Users gain the following advantages:
- The user quickly gets a general idea of the process-oriented application of the standards' requirements.
- The user can study which requirements have already been considered and which requirements have not been applied yet.
- Through the modification of neutral process modules the user can quickly develop company-specific process models.
- The model-based creation of QM documents is supported.

Process modules are integrated into the enterprise model in two fundamental steps:
1. After the identification of a QM element that has not been applied, users have to determine – on the basis of the selected system – the best position in the function model at which a process model should be embedded to fulfill the desired extension.
2. The second step contains selecting the process module of the library and adjusting the module to the corporate conditions. The adjustment includes designing the processes in consideration of the corporate process organization and giving activities, orders, products and resources the company-specific names.

The process modules that were created on the basis of the 20 elements refer to different subprocesses of the enterprise model. For the users of the model library it is therefore difficult to integrate the individual process modules into the suitable position of the enterprise model.

Users may therefore apply a predefined model that, e.g., gives a general idea of a possible arrangement of the QM process modules in the enterprise model. Through a target performance comparison the user can additionally determine QM elements that have not been considered yet in the user model. This ensures that all elements that are relevant to the QM system are indeed incorporated. The model is based on a performance-oriented arrangement of the QM process modules into process modules of control – process-related and process-accompanying – and the object-oriented assignment of the modules to the object classes 'order', 'product' and 'resource'.

With regard to the control-related QM process modules, users have to notice the following aspects:
1. The control-related QM process elements represent the long-term and periodic QM management tasks. They generate orders for all other processes.
2. They serve as a checklist for these activities.

The description of processes through the IEM methodology is based on the definition of objects (products, orders, resources) of the studied system and the organization into object classes. Therefore, to utilize process modules, there must be pre-structured class trees in an object class library that – among other things – contains the necessary objects of a QM system according to DIN EN ISO 9001. The QM object class library is based on the IEM reference class library that was enlarged by specific objects of the user view 'quality management' These objects are necessary specifically for the development of the user view 'quality management'. They can be used for the model-based creation of QM documents.

Quality-Oriented Design of Business Processes 63

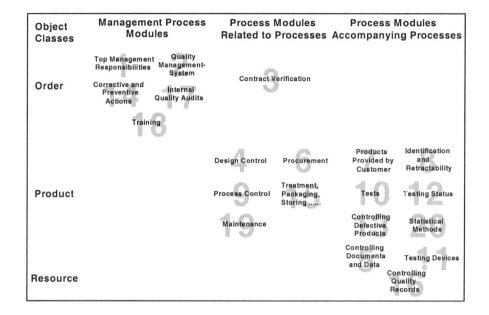

Figure 27: Assignment of QM Process Modules to the Object Classes

The performance-oriented organization of process modules is necessary because the utilization of process modules requires different rules of application (cf. chapter 4.4). Basic aspects of the use of process-related QM process modules include:
– The process-related process modules refer directly to the production of goods and should therefore be integrated first.
– They serve as orientation and as example.
– They have to be adjusted to the corporate situation in detail.

Basic aspects of process-accompanying QM process modules include:
– The process-accompanying QM process modules contain various interfaces with the processes of direct production. They should be used following the process-related process modules.
– The represent a quasi-standard concerning the application of the requirements.
– They require few adjustments to the corporate situation.

Examples of specific objects of quality management are documents that are created for the description of a QM system (QM manual, QM procedure instructions, QM work instructions). Each QM process module contains, e.g., the procedure instructions as a resource of the respective QM activity. These procedure instructions are defined in the object class tree of the generic object class 'resource' in a subclass 'quality-related document'.

When importing QM process modules into the company-specific model, the specific object classes of the user view 'quality management' are automatically added to the class libraries of the specific enterprise model.

The reference model for the performance-oriented system delimitation of the QM process modules provide users with a structure that they can use to integrate the process modules into the enterprise model. If the user attempts to model the QM system according to DIN EN ISO 9000 ff he can use the reference model and the QM process modules as a basis.

The activities are described by the features 'objects' (what is processed?) and 'performance' (what is done?). These features are neutral and can be taken over directly into the enterprise model. The features 'method' (how?), 'person responsible' (who?), 'resource' (with what?), and perhaps 'location' (where?) and 'time' (when?) are company-specific. These features have to be described by the user of the reference modules on the basis of the specific corporate situation. For the features 'person responsible' and 'resource' possible objects have already been defined in the resource view. These have however been marked with a question mark because the user still needs to adjust the features to the company-specific situation.

Chapter 4

Modeling Rules for Quality-Oriented Design of Business Processes

1. GOALS AND REQUIREMENTS OF MODELING RULES

The use of methods to model business processes or certain parts or aspects of companies (e.g., quality management) in light of an increase in the complexity of planning tasks is almost compelling. Models contribute decisively to the control of tasks and efficient and optimal results. The necessity of modeling is based on the increasing communication requirements of management, corporate planning, supporting and concerned areas. This requirement can be fulfilled by models that serve as a common basis for discussions. The purposive and efficient application of modeling methods is usually only possible if the participants have a wide range of experience with the used methods.

Therefore, the goal of the following modeling rules is to support the application of Integrated Enterprise modeling (cf. chapter 3.2) and the use of model libraries (cf. chapters 3.3 and 3.4) for efficient modeling. This is to enable a wide range of applicability and an efficient application of IEM that is transparent for all participants. The application of these modeling rules pursues the following goals:
- time-related goals, e.g., by way of describing purposive approaches,
- cost-related goals, e.g., by way of explaining specific data surveys,
- quality-related goals, e.g., by way of describing approaches to create models that conform to methods and to represent quality-related data and

– social goals, e.g., by way of explaining team-oriented approaches.

The modeling rules should be used to describe how to apply the constructs of IEM. The application of IEM should be supported by rules, particularly in the course of corporate planning projects.

1.1 General Requirements of Modeling Rules

In terms of attaining the best possible results, modeling rules should essentially meet the following requirements:
- High degree of transparency and clear description,
- structured documentation,
- best possible level of detail and
- neutrality.

A high degree of transparency and a description that is easy to understand are prerequisites for the ability to communicate modeling rules and for the application by inexperienced staff members. The best possible level of detail, i.e., the appropriate refinement of the approaches, prevent that the most important elements of the modeling rules are covered up by a complex description.

A logically structured documentation is necessary to utilize the rules during the modeling process as a source of reference. The application of modeling rules can be illustrated by examples. This would guarantee common understanding.

1.2 QM-typical Requirements of Modeling Rules

In the model-based development of a QM system, users may be faced with the following situations:
1. There is no model of the company available. The goal of modeling is the exclusive description or documentation of the QM system according to the standard DIN EN ISO 9000 ff.
2. There is no model of the company available. The goal of modeling is the creation of a complete enterprise model incorporating quality management.
3. There is a process-related model. The company would like to utilize the available models to document the QM system. Modeling exclusively focuses on documenting the QM system and is based on the 20 elements of the standard DIN EN ISO 9000 ff.
4. There is a process-related model. Quality-related aspects are to be added to the existing model. The goal of modeling is to develop a model – incorporating quality management – that can be used flexibly.

In reality, many companies try exclusively to develop a model that describes the corporate QM system. They focus on certification according to DIN EN ISO 9000 ff and would like to use the model to document the QM processes.

The goal of modeling should be to develop a model that can be used flexibly and in which specific aspects of individual views (e.g., quality management, environment management, information flow) can be incorporated. The rules must, therefore, enable the development of different views. This goes beyond the creation of a QM-oriented model. Due to the many different initial situations, users must be enabled to enter the modeling process flexibly. The rule should support this flexibility.

In addition – and with regard to quality management – the modeling rules should also provide rules that
- represent,
- identify possible weaknesses and
- plan and realize improvements (cf. [Kle94])

of typical tasks and processes of the corporate quality management.

Today, corporate quality management not only focuses anymore exclusively on the organization of measuring and testing. It rather concentrates on planning the design of all processes that are relevant to quality. This leads to the following fundamental requirements of quality management:
- to develop, document and describe QM systems,
- to design ('control') processes that are safer and more robust and
- to improve processes continuously.

Modeling rules should direct the user in such a way that he can utilize modeling purposefully for the above-mentioned tasks. Purposeful in the connection with the development of a QM system means that when creating, evaluating and modifying the models, the above-mentioned typical tasks of QM system design are supported by the modeling rules.

2. APPROACHES TO QUALITY-ORIENTED MODELING

2.1 Modeling Steps

On the basis of [LAN79] and [Fer91] the IEM modeling rules distinguish the following modeling stages:
1. system delimitation,

2. model creation,
3. model evaluation and utilization and
4. model modification.

In the stage of system delimitation the system (e.g. a corporate area) that is studied and modeled is selected and delimited. The user determines which areas are modeled on which level of detail. He determines which corporate areas, objects, organizational units, staff members, information systems and documents are included in the model.

Model creation is the developmental stage of models. The development of previously non-existent models to a status that is recognized as being correct and sufficient characterizes this stage. In corporate planning projects, users usually develop company-specific models to represent the actual state (actual state model, cf. example in chapter 7). If in the course of a project, users do not create actual state models the stage of model creation describes the description of target states in target models. On the basis of the two main views 'business process model' and 'information model' (cf. chapter 3.2) users can analyze for which aspects of the studied system they wish to develop additional sob-models.

In this stage the users select the type of model and the model design (process modeling), the data modeling, the design of the output and the model test.

The model creation in the above-mentioned stages results in a documentation of
- object hierarchies or information models that contain structured descriptions of the objects,
- process models that contain descriptions of the functions and processes that modify the objects and lists of the authorization and the executing resources and
- further sub-models that are developed from the business process and information models and that represent specific aspects of the objects.

Figure 28 illustrates the desired results of the first two stages regarding the information- and function-related description.

In the stage of model evaluation users identify weaknesses of the created models and estimate improvement potentials. Weaknesses can only be identified as such if the users have a (rough) idea of how to improve or optimize the studied system. Improvement and utilization potentials can only be identified with regard to targets. The model is used as a discussion basis or by way of the application of reference models to develop target concepts.

In the stage of model modification existing models are used to efficiently create modified models to document planned target states. These again can ultimately be used to describe the new actual state. The use of existing models is usually much easier and less expensive than a new model of the target concept. Existing models are thus modified. These changes are considerably facilitated by computer-aided modeling tools (cf. chapter 5).

Models are usually modified by way of updating actual state models in order to create target models (examples cf. chapter 7). Users can also modify models in company-specific adjustments of neutral reference models (cf. chapter 4.4) or when transferring existing models from one company to another similar one.

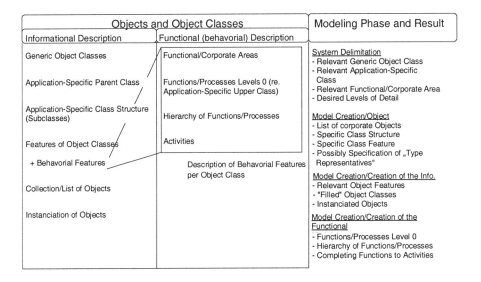

Figure 28: Functional and Information-Related Object Descriptions in the Stages

The following chapter contains specific and detailed approaches or modeling rules for all four above-mentioned modeling stages.

2.2 Ways to Quality-Oriented Modeling

As described in chapter 5.1.2, there can be different situations at the beginning of a quality-oriented modeling process. In some cases, there is a model to which only the user view 'quality management' needs to be added. The goal of modeling should be to develop a flexible model that goes beyond describing the QM processes on the basis of the 20 elements of the standard DIN EN ISO 9000 ff. It should be suitable for the development of further views.

Quality management systems have to be developed, then secured with suitable methods and finally continuously improved (cf. [Klei94]). The fulfillment of these requirements results in the following tasks:
- the specification of QM elements to be realized and their application within corporate processes,
- the identification of critical processes/activities and the specification of protective and
- the specification of quality-relevant variables as well as their measurement and evaluation in the processes.

Therefore, the following statements refer to the basic requirements or steps of the development of the user view 'quality management'. The realization of above-mentioned and other tasks of quality management is supported efficiently by the application of models. Table 1 illustrates the modeling tasks that are necessary to apply the above-mentioned requirements of quality management.

Basic Requirements of Quality Management	Modeling Task / Stage
Creating QM Systems	Modeling sub-processes that correspond to the conversion of QM elements (e.g., from a model library)
Controlling Processes	Improving critical processes / activities through protective control circuits
Improving Processes continuously	Modeling evaluation methods to determine quality-relevant indicators

Table 1: Quality Management - Basic Requirements and Modeling Tasks

In the first step, sub-processes are modeled that correspond to the corporate application of the requirements of the elements of DIN EN ISO 9000 ff. In this step, the QM process modules of the model library are used. The sub-models are incorporated into the entire model at suitable positions – if necessary, in consideration of the sample model to integrate QM process elements (cf. 3.4.3).

In the second step critical activities or processes are identified. They are made safer by a control circuit that needs to be specified and integrated. To

Modeling Rules for Quality-Oriented Design of Business Processes 71

continuously improve the process the third step contains the insertion of data collection activities at suitable locations within the process model and the integration of processes to evaluate the data. The indicators relevant to quality (e.g., the number of faulty deliveries) can be used to improve a subprocess or the process networking (e.g., adjustment of the testing level). Table 2 illustrates the assignment of tasks of quality-oriented enterprise modeling to the modeling stages.

Quality-Oriented Modeling Modeling Stages	Application of the QM Elements of DINEN ISO 9000 ff and Documentation in Sub-Models	Ensuring Processes through Modeling and the Creation of Control Circuits	Modeling Quality-Relevant Figures and Evaluation Methods
System Delimitation	Determining the relevant processes and object classes in consideration of the modeling goals	-	-
Model Creation	Integrating sub-models of the QM elements to be implemented into the process model	If necessary, specifying the data to be recorded (controlled variable); defining class attributes; determining sources for measuring figures	Specifying quality-relevant indicators; defining the included data as class attributes; determining sources for measuring figures
Model Evaluation	Identifying QM elements that were not realized	Identifying critical processes / activities that should be protected by control circuits; specifying the control circuits	Creating and evaluating the indicators to identify weaknesses and potentials and evaluating alternative (improved) processes
Model Modification	Integrating sub-models for the selected QM elements	Specifying and integrating the control circuit into the process model	Inserting the models for evaluation methods into the modified process model

Table 2: Tasks within the Modeling Stages in the Course of Extending Quality-Assuring Measures

3. GOAL FINDING AND SYSTEM DELIMITATION

3.1 Goals

The goal of system delimitation is the demand-oriented restriction of the area that is to be modeled to enable efficient modeling. The settings are based on the specific modeling task. The task also determines the relevance of the components of the system.
You can delimit a system in two main steps:
1. determining the limits of the model: the 'width' of the studied area of the system is determined.
2. determining the desired level of detail: how accurately, on how many levels of hierarchy should this area be modeled.

The results of system delimitation activities should not be regarded as definite borders. They can be modified, if necessary. Especially the desired level of detail can change: when modeling progresses or when identifying weaknesses and improvement potentials, users may require more details.

3.2 Determination of the Limits of a Model

The approach to restricting the system that is studied and that should be modeled can occur in two steps:
1. determining the modeling goals and
2. restricting the system that needs to be modeled according to the following criteria:
a) objects to be studied:
- identifying the relevant object classes (products and/or resources and/or orders) that are to be studied.
- determining the most important object class, a so-called 'primary object class', as focus of attention (usually the class 'product').
- identifying structures: determining which subclasses/objects of this most important class should be studied.
- developing/assigning which subclasses/objects of the two other classes should be studied.
b) functions to be studied:
- identifying for which object classes you need to describe changing functions.
- determining which object-modifying functional areas, processes and functions for products, orders and resources should be studied. If you have determined a 'primary object class', you should initially determine the functional area for this object class.

c) times to be studied
- determining the times that should be studied (e.g., night work); you should only model those object changes that occur during these times.
d) locations to be studied
- determining the locations to be studied (e.g., a certain factory building); you should only model those model changes that occur at these locations.

The criteria to delimit the studied system should be used in consideration of the modeling goals. They complement each other and can be applied in any sequence.

Users may also delimit the system according to additional criteria. Some corporate planning projects particularly delimit organizational units as specific object class restrictions. Actually, you should usually not determine the organizational units to be studied. Insisting on grown corporate structures should, if possible, be avoided.

Usually, the modeling task will determine which criteria should be applied first. Other criteria can then be applied additionally to restrict the modeling range further and to reduce the modeling expenses.

However, on principle you should follow the following rule: The borders of the system should be as narrow as possible, but as wide as necessary. This means that you should use as many criteria as possible to delimit the system that is studied.

To integrate quality management into the description of the model it is required to include QM aspects when delimiting the system. This concerns the steps 'restricting the objects' and 'restricting the functions'. Both steps are oriented at the requirements of the international standard DIN EN ISO 9000 ff.

3.3 Determination of the Desired Level of Detail

The goal is to determine a limited number of hierarchy levels of objects and functions. This will enable you to collect information and create the model purposefully.

To determine the level of detail you have to indicate name those levels of hierarchies of the studied objects and functions that should be modeled. Clear reference figures can neither be given for the hierarchy levels of the functional hierarchy nor for the hierarchy levels of the objects' hierarchy. For the description of the desired hierarchy levels you should, therefore, create examples of the highest and lowest hierarchy level of objects and functions.

On principle, you should follow the following rule: There should be as few levels of detail as possible and as many as necessary.

Just as the system delimitation, the level of detail depends on the respective modeling task. The set level of detail should not be viewed as a final setting, but should be adapted as modeling progresses. You should, therefore, determine a 'desired' level of detail.

To determine the level of detail you can use the criteria
- object hierarchy and
- functional hierarchy.

You should apply that criterion first that you have already preferred when delimiting the model.

4. MODEL CREATION

4.1 Goals and Desired Results

The goal of model creation is to represent the delimited area transparently by way of creating original models. With regard to IEM, this stage focuses on the description of the two main views with constructs of IEM. You should distinguish the stage of model creation, in which new models are created, from the stage of model modification, in which existing models are changed or modified. The stage of model creation ends if – for the respective application – the created models represent the relevant sections of the system as complete as possible, in detail and correctly or if the modeling goals of the respective application are attained.

To apply the modeling concept of IEM you should follow the four following main steps of model creation according to the predefined generic object classes and main views:
1. Identification of the object classes 'product', 'order' and 'resource' that should be modeled within the set borders of the system.
2. Creation of process models, i.e., identification and description of activities and processes.
3. Creation of information models.
4. Development of specific sub-models to represent aspects that are relevant to planning and/or modeling.

All models of IEM represent certain aspects of certain objects. According to the object-oriented approach of IEM, these objects are always subclasses of the classes 'product', 'order' or 'resource'. For the section of the system that was restricted in stage 1, the stage of model creation leads to the following results:
- Company-specific (or task-specific) specification of the object classes 'product', 'order' and 'resource',

- specification of subclasses and descriptive attributes,
- a database that is structured according to the object classes and their attributes and that describes the objects,
- if relevant for the respective modeling task, a hierarchical description of object-modifying functions/activities, their processes, controlling orders and executive resources,
- models that represent aspects that are relevant for planning or decision-making, e.g., a Gantt diagram to represent the execution time of functions and functional processes.

4.2 Identification of Objects and Object Classes to be Modeled

The goal of object identification is to identify and structure all objects and sub-objects that are changed in a delimited system. At first, you need to verify those object classes that were identified in the stage of system delimitation. If necessary, they have to be specified in detail. You should also specify those class-specific attributes that describe the object classes. If required for the specific application, you can then identify and clearly name the objects of the object classes.

The stage of object identification results in a description of the user-specific object classes and class structure. The descriptive attributes of the object classes are provided and the objects that are relevant to planning within the borders of the system are listed and assigned to the classes.

The identification and organization of objects in object classes are part of the information-oriented description within IEM. Figures 29 illustrates in which modeling stages or steps the information-oriented object description occurs. In the stage of system delimitation the relevant object classes are determined. If required, this stage also includes the creation of corresponding object classes of the highest user-specific level. During the stage of object identification the class structures are modeled and the classes are specified by detailed class attributes. To create an information model based on these two steps you now have to instantiate the objects.

4.3 Creation of Process Models

The goal of process modeling is to analyze and understandably represent corporate business processes. The processes should be recorded independently from the organizational structure and be described in a department-overlapping way. This will illustrate the coordination and cooperation activities of the corporate departments as well as weaknesses and it may motivate to force open departmental principalities.

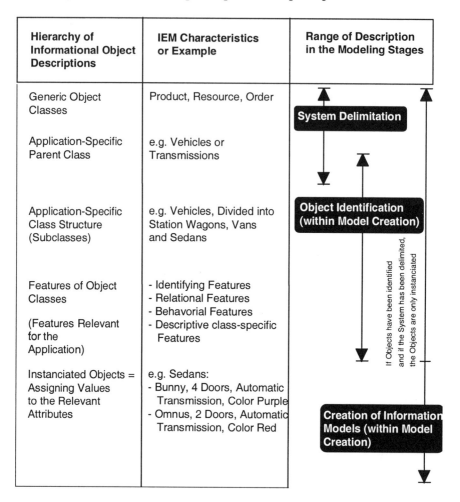

Figure 29: Stages of Descriptions of Objects and Object Classes

Process modeling contains the following steps:
1. Identification of functions that modify products or the 'primary object class';
2. Delimitation of the functions;
3. Connection of the functions to logical processes;
4. Specification of the functions and processes top-down or summary of functions and processes bottom-up;
5. Assignment of executive resources and initiating orders to the functions

4.4 Creating QM-Related Process Models

4.4.1 QM Process Models

To support the creation of QM systems according to the international standard DIN EN ISO 9000 ff users are supplied with QM process modules (based on DIN EN ISO 9001) in the form of reference models. These can be transferred into the functional model (cf. chapters 3.3.3 and 3.3.3).

General Procedure:
1. determining the required QM process module,
2. getting a general idea of the illustrated realization of the requirements of the element,
3. assigning the QM process module to the respective hierarchy level of the model,
4. integrating the QM process module into the respective process model,
5. adapting the objects (products, orders and resources) to the specific corporate conditions.

Detailed Procedure:

In the first step you have to determine the process module that should be implemented. This module has to be integrated from the model library (cf. chapters 3.3.3 and 3.4.3) into the user model. The class trees of the QM process modules are based on the predefined class tree of the IEM methodology. Aspects of quality management were added to the class tree.

The QM process modules contain a predefined part of a process model that takes all requirements of the international standard DIN EN ISO 9001 into consideration. Users that are not familiar with this standard should at first try to get a general idea of the process that is represented. When he understands this idealized process he can realize the company-specific adjustments.

The QM process modules have to be integrated into the respective hierarchy level of the model. Especially in the case of an existing model in which the stage of system delimitation did not focus on the introduction of QM, the QM process modules have to be integrated into the different levels.

In this step you should first determine at which point of the existing subprocess you want to insert the QM process modules. Afterwards, the connections between the affected activities are disengaged, the QM process modules are inserted and are then connected with the preceding and following activities.

The QM process modules stored in a reference library represent idealized processes that have to be adapted to the real company. At first you have to adapt the names of the actions to the language that is used in the company.

If, for example, the QM process module calls an action 'examining a contract', the internal communication, however, refers to each contract as a 'customer order', the action has to be renamed. This also applies to the other object classes 'products', 'orders' and 'resources'.

Afterwards, the assignment of individual activities in the model to the respective QM element according to DIN EN ISO 9001 is documented. For this purpose, a resource of the resource class 'QM procedure instruction', that refers to the respective QM element, is added to each activity. For example, the resource 'QM procedure instruction contract examination' is assigned to the activity 'examining a contract'. On the one hand, this illustrates that the activity requires the valid procedure instruction; on the other hand, it documents the assignment of an activity to a QM element (cf. chapter 6.3).

Finally, you can define attributes for the individual object classes. Attributes of the resource class 'QM procedure instruction' may include
- 'realization' (value: completely, partly, not realized) and
- 'examined' (value: scheduled, effective, will be modified).

Attributes that are used often (e.g., attributes that are used for the model-based creation of manuals) have already been defined in the QM process modules.

For the process-accompanying and the control-related QM process modules (cf. chapters 3.3.3 and 3.4.3), you can give an abbreviated procedural description since a detailed adjustment to the user model is not necessarily required. The following procedure applies to both cases:
1. determining the necessary QM process module,
2. getting a general ideas of the tasks that should be carried out,
3.   assigning the QM process module to the respective hierarchy level of the model,
4. adapting the objects (products, orders, actions and resources) to the specific corporate conditions.

Detailed Procedure:

Steps 1 and 2 have already been described for the process-related QM process modules.

In the third step, the QM process modules are integrated into a suitable hierarchy level of the model. Chapter 3.3.3 contained a possible structure for the implementation of the QM process modules. This structure was based on the structure of the standard DIN EN ISO 9001. In general, the process-accompanying QM process modules are arranged parallel to the process-related process modules. They are modeled in the product view. The control elements generate orders for all other QM process modules and are modeled in the order view. In the fourth step, the objects are adapted to the specific conditions – just as in case of the process-related process modules. In

Modeling Rules for Quality-Oriented Design of Business Processes 79

contrast to the process-related QM process modules there will only be few adjustments.

4.4.2 Models of Control Circuits

To stabilize and control corporate activities or processes in the sense of quality management, there are models of control circuits that are supplied as reference models and that users can incorporate into the process model at critical positions. Apart from predefined models of control circuits, control circuits can be used to represent the applied procedures and methods of quality management (e.g., SPC, QFD, K-FMEA).

General Procedure:
1. determining criteria and data to identify critical activities and processes,
2. identifying critical activities and processes,
3. selecting a control circuit to stabilize the critical process,
4. integrating the control circuit into the user model and adapting the control circuit to corporate conditions.

Detailed Procedure

To identify critical activities and processes the user has to determine criteria that illustrate the qualitative ability of the company. Possible criteria include:
- high error rate,
- many complaints,
- high expenses required to identify and eliminate errors or
- results of analytical studies, e.g., process failure mode and effects analysis.

The above-mentioned criteria can be quantified with suitable data. The recording and processing of suitable data can be modeled through the application of process models for data acquisition.

On the basis of these criteria you have to determine those activities or processes that effect the company negatively or are not controlled sufficiently. The integration of control circuits is always fairly expensive and increases the complexity of the corporate processes. Through cost effectiveness analyses, you should, therefore, identify those activities or subprocesses that contain high potentials for improvements.

Not each activity or process can be controlled on a short-term basis. An example of such a process is the attempt to guarantee on-time deliveries. This process contains many risks and cannot be influenced by predefined measures. You should select activities or processes
- for which you can indicate defined control and target figures (quality features),
- whose control and target figures can be measured,

- which can be influenced reproducibly by specific quality assurance measures and
- that can be secured through the participation of additional positions.

Critical processes should be characterized by suitable attributes (attributes: critical process; value: identified, protection planned, secured).

To secure the working results of critical processes you should determine measures and describe these within a control circuit. You can use the predefined control circuit modules that are supplied in the model library 'quality management' (cf. chapters 3.3.3 and 3.4.3). Apart from these predefined control circuits that are based on models of traditional control system engineering, the model library contains specific control circuit modules that represent procedures and methods of quality management. Examples of such procedures are Statistical Process Control (SPC), Quality Function Deployment (QFD), Design Failure Mode and Effects Analysis (K-FMEA) and Self-Evaluation.

The control circuit has to be inserted into the hierarchy level of the critical activity. The adjustment of the control circuit to the corporate conditions is based on the QM process models and occurs in the following steps:
1. using internal language to name the objects,
2. adding missing objects to the control circuits,
3. defining attributes for the individual object classes.

4.4.3 Process Models for the Derivation of Indicators

The collection and processing of quality-relevant indicators or data are tasks that are immensely important for the continuous improvement of processes in the sense of quality management. Examples of possible indicators include:
- indicators that enable the protection or improvement of corporate processes, e.g.
- indicators to evaluate suppliers,
- indicators to examine contracts,
- indicators to control processes,
- indicators to determine the cost of errors.
- indicators that are required to hedge liability risks, e.g.
- indicators of product tests,
- indicators of model tests,
- indicators required by customers.

The integration of data collection activities and the design of processes to determine indicators create a basis on which specific improvements of processes (e.g., reduction of testing expenses) can directly be applied.

Modeling Rules for Quality-Oriented Design of Business Processes 81

General Procedure:
1. determining the required quality-oriented indicators and data,
2. determining the data that has to be collected and assigning attributes to classes,
3. representing the evaluation of the data in a separate model,
4. inserting the activities that are necessary to collect the data into the respective models.

Detailed Procedure:

You have to identify those indicators of which you expect that they will yield findings of quality-oriented processes, reduce liability risks or have been asked for by your customers. Examples of possible indicators were given in the previous chapter. When determining indicators, you should consider the following aspect:

– Only measure what you also document.
– Only document what you also evaluate.
– Only evaluate what has informative value.

In the second step, you should determine which data are required to generate the indicators. In consideration of the expected utility potentials you should analyze whether the required data can reasonably be incorporated. The data should then be assigned to suitable classes and be specified by attributes (e.g., number, type, and time and frequency of measurements).

In consideration of the aspect described in step 1 you should initially use a sub-model to represent how to evaluate the data. To begin with, the source of the data is irrelevant. This will be determined in the following step. The sub-model should be arranged on a suitable hierarchy level of the entire model.

In the model, the sources of the data required for the evaluation should be marked by way of assigning monitoring stations. Monitoring stations are represented as actions. In the case of complex measurements you should specify the action. The determined data should then be assigned to a specific product, order or resource class. It is always possible that monitoring stations are at completely different positions and on completely different levels of the mode!.

4.5   Creation of Information Models

The goal of the information model is to organize the corporate data in a structured directory and to design the supply of data for the execution of all subtasks within the company [WAL89]. The information models are created at the same time that the processes are modeled.

When creating an information model the identified objects have to be described by attributes that have previously been determined for each object class. The characterization of general descriptive features by specific attributes (attribute values) of specific objects is called 'instanciation'.

The following steps are required to develop information models:
1. determination of the generic object classes that should be modeled and determination of the highest user-specific classes,
2. identification of the class structures (subclasses) and the objects that should be modeled and determination of the respective (class-specific) attributes that describe objects (4.4.2),
3. selection of attributes that are relevant to planning tasks and that can be characterized by values,
4. instanciation of objects, i.e., characterizing the relevant attributes through specific values,
5. iteration: during steps 2 through 4 you can return to the preceding step at any time to specify or supplement certain results.

As illustrated in Figure 29, the modeling steps of system delimitation and object identification are seen as components of the procedure to describe the IEM object classes, i.e., to create an IEM information model. After the system has been delimited and after the objects have been identified the object classes, that should be modeled, their descriptive attributes and the relevant objects of these classes are already known. To create the information model you thus have to select the relevant attributes. For all objects you will then have to assign values to these attributes. At this point, it may become obvious that you need to introduce additional object classes or descriptive attributes.

4.6   Creating QM-Related Information Models

QM-related information models refer to the data that is required to develop the user view 'quality management'. An example of QM-specific data is information that is recorded to determine indicators in the course of the evaluation and improvement of measures to assure the quality. For this, you need to identify data that is relevant to quality. This data needs to be measured and described by indicators.

The indicators should be specified as attributes of an object class. You should
- determine those indicators that should be used to evaluate processes or task results,
-   specify the necessary data, i.e., the data that was used for the indicators and
- assign this data to an object class.

The necessary attributes are assigned to object classes by way of

- determining the generic IEM object class that is described by the feature and
- examining the appropriate company-specific class tree top-down according to the class for which the feature acts as descriptive feature.

For the determined indicators you can indicate locations in the process model at which the attributes of these indicators should be measured. The results of these measurements then determine the further process.

4.7 Development of Specific Submodels

At first you should check which submodels should be developed (first determination before creating the process and information models). You should then identify those attributes of the information and process models that are to be emphasized and represented in a submodel. You should also check whether these attributes are complete. In a third step, you have to transfer the attributes – for an area that needs to be determined – into a form that is predetermined by the submodel or the method to develop submodels (e.g., a Gantt diagram).

5. MODEL EVALUATION

5.1 Targets and Possibilities of a Model Evaluation

The goal of a model evaluation is to utilize the created models in such a way that relevant statements can be made about the task that was the initial basis of the modeling process. The created models are thus utilized according to the modeling task. In the course of corporate planning projects the models serve as a common basis for discussions and as a common, complete and 'correct' view onto the studied area of the company. Concerning the achievement of corporate objectives, they also help users to identify weaknesses and the resulting potentials for improvement. Therefore, the utilization and evaluation of models is necessary to develop measures that correspond to the true needs of the company.

In the example of corporate planning tasks, the main steps of evaluating IEM models include
- the identification of weaknesses and
- the estimation and assessment of improvement potentials.
  The results that should be achieved include

- a common and uniform understanding of all participants of the objects and processes that were modeled,
- a list and a specification of all corporate weaknesses within the studied section – including the effects – and
- a list, evaluation and prioritization of improvement potentials.

5.2　Specification of Improvement Potentials in Quality Management

The use of IEM models enables users to identify weaknesses and develop improvements that increase the ability for quality of the company on a long-term basis. In this context, we should advise you that the application of the developed improvements in the corporate reality is the decisive step to increase the quality. Staff members should not only commonly know how the modeled objects and processes are structured: they should 'live' the system of quality management. It is, therefore, necessary that the potentials of quality management are explained and illustrated to staff members and decision-makers. This requirement can effectively be realized through a target performance comparison and a model of the company. The target performance comparison should be oriented towards the development steps of quality management.

The design of quality management takes place in a cycle that contains three steps:
1. setting targets for corporate processes,
2. complying with the targets of corporate processes and
3. increasing the targets for corporate processes continuously.

These steps are supported by the following aids:
- process modules for quality management (cf. chapters 4.3.3 and 4.4.3),
- control circuit models (cf. chapters 4.3.3 and 4.4.3) and
- process models to derive indicators (cf. chapter 5.4.4).

In the following we will present which quality-oriented improvement potentials can be developed from the application of the above-mentioned aids, how to represent these potentials and which advantages users can expect.

The process modules of quality management supply predefined parts of processes on the basis of the requirements of the standard DIN EN ISO 9000. In the first step, the QM modules can be used to create and describe the company-specific QM modules.

The utilization of the modules enables the user to estimate the expenses for introducing and documenting a QM system according to DIN EN ISO 9000 ff. In the course of a target performance comparison you have to determine those elements or requirements that have so far not been fulfilled.

Modeling Rules for Quality-Oriented Design of Business Processes   85

You can support and document this comparison if you supply suitable attribute values for the objects of the class 'resource'. The documentation of the comparison additionally allows you to trace the efforts made to develop the QM system. The company-specific QM modules that were developed while creating the QM system are the basis for further estimates of potential improvements.

Further aspects of the support of the identification of weaknesses and potentials through QM process modules include:
- The time needed to model the processes to represent the QM elements is reduced because users primarily need to adjust certain aspects. Users have more time to identify weaknesses.
- The security increases because the model now contains all relevant QM elements.
- The modeler does not have to newly develop the QM modules because the system proposes the basic processes.

In the second step, we have to ensure that the predetermined QM requirements are fulfilled. To control corporate activities and processes you have to determine – through suitable criteria – and document in the model those activities or processes that effect the quality of the company decisively and that in the past were frequent sources of errors. The quantification of the criteria reveals whether the expenses that would be required to introduce additional control circuits (or traditional measures and methods of quality management) can be justified economically. The identified weaknesses concerning production processes and the corporate process organization can then be reduced through the introduction of control circuits. The introduction of control circuits leads to avoid errors. The utilization potential is therefore especially large for those processes that contain many errors.

Further utilization potentials of the introduction of QM control circuits regarding the estimates of potentials include:
- The supply of a basic element, the control circuit, facilitates the target concept of a model that is the basis for estimates of potentials.
- Due to its structure, the control circuit cannot only act locally, but throughout several levels of the model. Estimates of potentials are therefore not restricted to one level.
- The supply of sample control circuits ensures that control circuits are modeled completely. Activities or resources that are important to estimate potentials cannot be overlooked.
- Typical procedures and methods of quality management can be integrated into the enterprise model quickly. Advantages and disadvantages of the introduction of procedures can be estimated.

In the third step, users should increase or improve the predetermined and controlled requirements on the basis of suitable indicators continually. The

description of evaluation methods documents the registration of indicators that are used to identify weaknesses and to estimate potentials. Utilizing these indicators you can make qualitative statements regarding the utilization potentials of new requirements. This supports decision-making processes. The determination of suitable indicators is sensible if the characteristics of the identified weakness are unclear. You should distinguish between the
- evaluation of indicators from the process (e.g., statistical process control in a manufacturing plant) and the
- evaluation of indicators to evaluate the effectiveness of organizational measures in the process model (e.g., before-and-after comparison following the reorganization of offers processing).

Additionally, the advantages of process models to derive indicators include:
- The process of evaluation methods to identify weaknesses and to estimate potentials can be represented in a specific submodel.
- Through an additional activity the sources of measuring data in the model are immediately identifiable.
- The submodel provides input for later simulations to identify dynamic weaknesses.

The presented aids support the specification of improvement potentials in quality management. The representation of the improvement potentials is the basis of decision-making concerning a change of corporate processes. It also contributes to a higher acceptance on the part of affected staff members.

6. MODEL MODIFICATION

6.1 Targets and Desired Results

Company-specific models are modified when creating models cyclical, when developing and converting target concepts and when expanding the studied section of the company. The utilization of existing models is usually much easier and less expensive than the development of new models. Existing models are thus modified.

Model modification contains the following three main steps:
1. determining the base model,
2. identifying the demand for modifications and
3. executing the necessary operations.

The goal is to efficiently model current and consistent enterprise models utilizing existing, i.e., documented and evaluated, modeling and planning know-how (e.g., actual state models, reference models).

Model modification leads to current models that enable the user to find the optimal solutions of the respective problem. Examples of such problems include:
- the reorganization of parts of a company or the entire company,
- the expansion of the company – either with the intention to increase the production or with the intention to expand the product range,
- the optimization of the corporate processes,
- product studies (determination of core processes) and
- the reaction to changing market requirements.

6.2   Determining the Base Model

The selection of the base model effects the quality and efficiency of model development. Users should consider the following aspects when selecting the appropriate model:
- The existing models of the company, e.g., from the actual state analysis, are usually the basis of updates (e.g., in the course of developing target concepts).
- You should check whether there is a reference model for the specific application and the type of company you wish to model.
- You should check whether you can utilize models from other projects or from other (similar) companies – apart from reference models.

There may be several base models. You should select the model that requires the lowest modeling expenses and that suit the specific modeling task best. A good selection usually depends primarily on the intuition, the knowledge and the experience of the modeler.

6.3   Identifying the Modification Requirements of an Existing Model

There are five possibilities for model modification. These may occur in any sequence in the course of a modification process. All requirements are characterized by the number and complexity of the following possibilities of model modification:
1. Refining an existing model to represent parts of a company in detail,
2. supplementing an existing model to represent additional parts of a company that have so far not been illustrated,
3. removing parts of a model that are no longer current or needed,
4. modifying parts of an existing model and
5. summarizing corporate areas, that were modeled in different models, to develop a complete and uniform enterprise model.

6.4 Executing the Necessary Operations

The changes occur according to the representation of the model. A computer-based tool that has been tailored to the method of IEM supports the described tasks and enables users to modify models easily and consistently.

Chapter 5

Computer-Based Tool

1. SITUATION, GOALS AND REQUIREMENTS

To design business processes it is necessary to collect information on the quality of processes, on whether and how processes can be controlled, on responsibilities, process optimization, process costs, environmental compatibility and on comparisons of processes (process benchmarking). The required data and structures and the models of the organizational structure and the process organization are often alike or similar. A computer-based tool for business process modeling and design must take these different views into consideration. It has to enable users to utilize the collected data and the models over and beyond the immediate task.

The goal is to use the business process models continuously and uniformly as a consistent basis of information. However, the modeling expenses should not become priceless. Modeling should occur task-related, i.e., you should only model the information that is really necessary to fulfill the particular task. Apart from these requirements the modeling tool should also be flexible: One should be able to improve it. This enables users to utilize the once modeled information in follow-up projects with other tasks. Additional information is simply added to the model.

This chapter describes the structure, interface and functionalities of a tool that fulfills the requirements of business process design of chapter 2, that supports the modeling language, the reference models and the model libraries of chapter 3 and that supports the modeling rules of chapter 4.

The study of different modeling and CASE tools as well as the comparison of different modeling methods from SADT [Spu94] through

IDEF [Spu94] to object-oriented approaches from Rumbaugh [Rum91] and Coad/Yourdon [Coa91] supplied the basis for the development of a software tool to support 'Integrated Enterprise Modeling' (cf. chapters 3.2 and 5.2) [Spu94].

Practical experiences with modeling tasks in industrial projects [Joc95a] and studies in the course of the QCIM project provided the requirements' profile of the software tool. The development of the tool was accompanied by continuous application and validation of the respective prototypes in application projects. In chapters 6 and 7 we have included examples of the application of the tool.

2. TOOL CONCEPT

An important aspect of the IEM methodology (cf. chapter 3.2) is the integrated representation of all relevant modeling aspects of a company within one consistent enterprise model. Users can conduct simulation studies and analyses on the basis of this integrated model. This was achieved through the integrated representation of an object-oriented enterprise model for which we defined different views to create and process the model.

Correspondingly, the tool contains a data storage device for all relevant data of the model. The data is organized in classes, objects and object states. The core of the tool ensures the consistency of the class and object data – including their relations. Through the core users can access all model information. This information is represented by views onto the enterprise model (Figure 30). The model is created through the views 'business process model' and 'information model' (cf. chapter 3.2). Specific user views – especially those concerning quality – were integrated (cf. chapter 6). Predefined submodels can be stored as libraries. The concept also allows for additional independent views onto the enterprise model. According to IEM, the information is stored in the model core. The views contain results of analyses (e.g., diagrams of costs of times) or information that was created through interfaces with other software tools. Classes and attributes (features) can be defined and inherited freely. This ensures the flexibility of the views. To conform with the existing standardization efforts we have planned to represent parts of models and entire models according to STEP/EXPRESS (cf. Annex B).

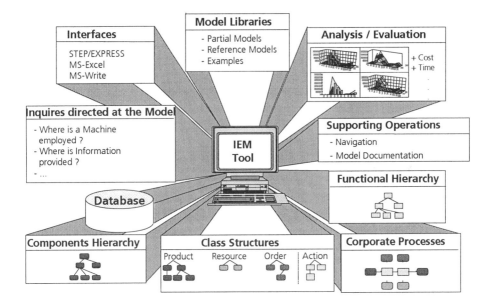

Figure 30: Tool Concept of IEM Tool (MOOGO)

3. THE ENTERPRISE MODEL IN THE TOOL

3.1 Modeling of Views

The tool, called MOOGO supports the modeling of the different views of an IEM model. The user only models the relevant sections of the views. Here, the tool ensures the consistency of the model. Modeling expenses are reduced through predefined structures (e.g., 'product', 'order', 'resource') and are restricted to the elements that are necessary to present the information. The context of the processes does not become lost. This method enables users to enlarge the model step-by-step [Joc94, Mer95e].

The tool secures the relations between the different views. These views are connected through the class structure. When describing a process you describe a section of the life cycle of an object that is involved in the process. All objects of the processes are instances of classes that have been defined in advance. Relations between classes can later be changed in the course of model expansions or modifications. Instances of different classes can be summarized on a higher level of abstraction into one instance of the common parent class (abstraction). Components of objects are determined as features of a class. The components are represented by the respective class (Figure 31).

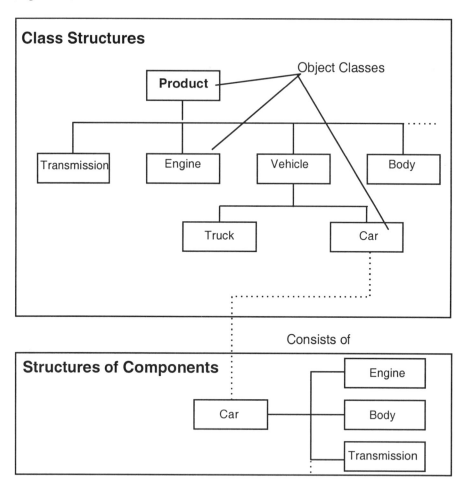

Figure 31: Connection of Classes and Components

3.2 Parameters

The object-oriented approach of IEM allows users to 'parametrize' the model – through a description of classes and attributes. Different values for attributes can be set and analyzed. The tool enables users to define and expend classes and attributes of existing models freely. It also provides mechanisms to import components of other models. This allows users to adopt a new view onto the model – simply by importing predefined class structures. For example, a model might have been created focusing on process analysis. Simply by importing the appropriate classes it can easily be utilized to solve problems regarding quality or environmental management.

3.3   Integration of STEP/EXPRESS

In the future, the exchange of data between different software systems will become increasingly important. In addition, the business processes of different companies have to be coordinated. For these purposes, we need to develop mechanisms that allow us to exchange data between the different systems that are used. STEP is a 'step' in this direction [EM91, ISO10303-49]. The description of product and process information in STEP is based on the standardized information modeling language EXPRESS [ISO10303-11]. The EXPRESS interface of this tool allows users to exchange model information with a STEP information model through a mapping mechanism (Figure 32).

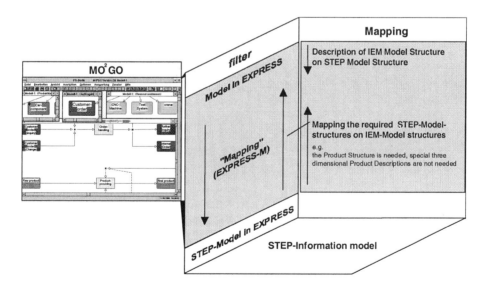

Figure 32: Concept of EXPRESS Interfaces

Further studies concerning descriptions of interfaces, especially with regard to workflow management systems, are carried out, e.g., by the Workflow Management Coalition [WMC95].

4. STRUCTURE OF THE TOOL

4.1 Modular Structure

We have developed an application-oriented modular concept that guarantees a high degree of independence of the individual modules in order to apply the concept of the tool to software engineering. The further development, changeability and maintenance of the tool are thus facilitated. .

The modular concept contains several levels. The highest level consists of the basic component of the tool and additional application-specific independent modules. The basic component is an autonomous program that contains all functions that are necessary for modeling and analysis. It also contains an interpreter for the additional user-specific modules. These are written in a macro language that is independent of the basic component. They allow implementation without comprehensive programming know-how. Users can establish a connection to other software systems through these modules.

The basic component can be divided into user interface, tool core and tool interface (Figure 33). The user interface has been organized into different classes that realize the dialog with the user. Graphic editors for classes and business processes belong in this category. The tool core is divided into further modules to manage and realize the IEM constructs. Each of these modules consists of classes that realize the respective task of the module. The interface of the tool core ensures a high degree of independence of the user interface

The rigid interface between the user interface and the tool core was established due to the following reasons:
- The display of the user interface (icons, windows, menus) can be changed without affecting the tool core;
- there is an interface for external users realized through specific procedures that do not affect the user interface;
- the interface tool can be exchanged and the portability between different platforms is ensured;
- different developers with different goals and responsibilities can work on the user interface and the tool core at the same time.

4.2 Hardware and Software Requirements

In practical application we have found that using Windows 95™ or Windows NT™ with a Pentium chip computer and 32 MB RAM accelerates things considerably. Especially when working with models that require more space than 10 MB (ca. 2.000 objects), such a computer is inevitable.

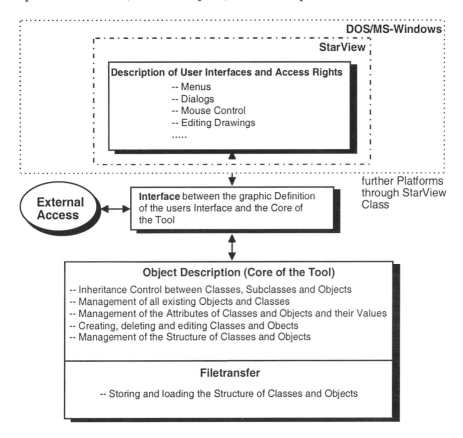

Figure 33: Modular Structure of the Basis of the IEM Tool (MOOGO)

4.3 Data Model

To transfer the tool concept into an executable program consistently you need an appropriate data model. A data model describes the required data and their connections that are necessary to execute a program. To conform with international standardization efforts we have chosen the language EXPRESS [ISO10303-11] to specify the data model. This facilitates the later representation of model information in standards such as STEP [ISO10303-1]. The specification of the IEM method with EXPRESS has been described in chapter 3.2.4.

5. USER INTERFACE AND FUNCTIONALITY

5.1 Modeling Components

The user interface of the tool was developed in practical application. The constant dialog with users ensured user-orientation. The cooperation of developers in projects has been successful. We were able to identify and specify the requirements much better. This approach resulted in a user interface based on Windows™. The editors needed for modeling are individual, application-specific windows in the sense of $MS_®$-Windows™, including the corresponding control functions (Figure 34).

The views that are necessary to model, analyze and design business processes are represented by different components of the tool (Figure 38). Examples of such components are class and business process editors and attribute and evaluation dialogs. They were integrated into the interface as graphic editors or dialogs. In the editors you can easily insert and 'manipulate' modeling elements with the mouse. All operations required to manipulate individual elements can be carried out directly through object menus. Global operations that concern several elements are activated through a menu bar.

5.2 Class Editors

The class editors correspond to the class structure of the IEM classes 'product', 'order' and 'resource' and their subclasses. The available commands are alike in all class editors. They enable users to create IEM classes and to add, delete and move attributes of a class. You can also change the class structure by way of moving classes through the class structure. The editors are used to describe and edit the different corporate objects.

5.3 Components' Editors

Within a class structure you can create structures of components for each class (Figure 34). The relations between the components are defined as features of a class. The components are always represented by their respective object class. It corresponds to the structure of objects of this class. Its direct components are assigned to a class. The hierarchy of components is created automatically for all components that have been assigned.

Computer-Based Tool

An example of applying the hierarchy of components is the representation of the organizational structure of a company. Further examples include the description of the composition of a product (e.g., bill of materials) or the structure of documents (e.g., QM manual, cf. chapter 7).

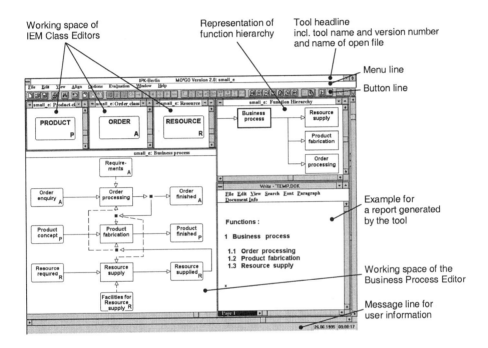

Figure 34: Components of the Tool Interface of MOOGO

5.4 Business Process Editor

The graphic representation of business processes is an essential part of an enterprise model. Each hierarchical level is represented in its window. The name in the top line of the window corresponds to the name of the senior process. The hierarchical description of the corporate processes serves to represent and analyze the business processes transparently (Figure 34).

There are no limits to the levels of detail. Copy and paste mechanisms allow users to model top-down or bottom-up. On all levels, modeling uses the same elements and follows the same criteria and rules. The tool thus ensures consistency between hierarchical levels and when representing business processes according to the logic of Integrated Enterprise Modeling.

5.5 Editing Attributes / Parameters

Business Processes and class definitions are the basis of numerous evaluations and analyses (cf. chapter 5. 7). To carry out specific evaluations you have to store additional information in the model.

Additionally, the tool allows you to supplement existing models with additional attributes for indicators or data. The indicators can be used along with the appropriate analyses to compare processes (Process Benchmarking or Process Structure Comparison). Additional data are used to generate manuals (quality management manuals, environment management manuals, organizational manuals) automatically.

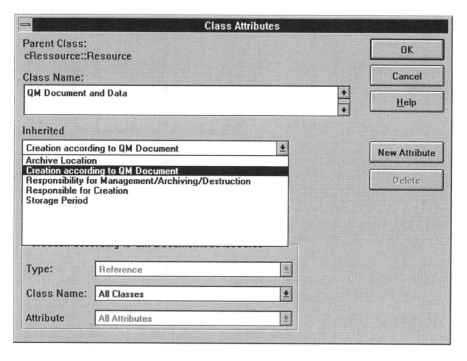

Figure 35: Attribute Dialog

The attributes are defined in the respective classes (Figure 35). The inheritance mechanism ensures that the attributes are available in all associated model elements and that they can be equipped with values. This results in a business process model that can be 'parametrized' (cf. chapter 6. 7).

5.6 Navigation in the Model

Business process models can easily get very large and complex. It is therefore very important that users can still navigate through and analyze such models. This includes search functions and extensive zoom mechanisms. Apart from the navigation through the model it is also important for users to be able to orient themselves in the printouts. This problem was resolved by way of numbering the processes and providing an index. This facilitates the work with the models.

6. INTERFACES

6.1 Interfaces with MS®-WINDOWS™ Applications

Important interfaces of each tool that supports the design of business process are those with word processing and spreadsheet analysis programs of Microsoft® Windows™. The are necessary to generate documents and to represent numerical evaluations graphically.

The connection to specific applications of MS® WINWORD™ and MS® EXCEL™ is based on a special macro language. For this purpose, a control sequence with model information and formatting instructions is described in this macro language of the tool. The sequence is understood by MS® WINWORD™ or MS® EXCEL™ macros. The respective application is then activated automatically.

6.2 Database Interfaces

All data and relations of the model are stored in a database. They are accessible through their SQL interfaces. At the moment, we use MS® ACCESS™. We have planned to install an interface with ORACLE.

6.3 Export / Import - Interfaces

Export and import mechanism allows users to create and load submodels. For this purpose, the tool uses a specific text-based language into which class descriptions, attributes and the graphics of the model are converted or in which they are represented. This language ensures the compatibility of models that have been created with different versions of the tool, the representation of models regardless of the platform, the evaluation or analysis of the model and the development of model libraries, reference class structures and open interfaces with other software tools.

7. EVALUATION MECHANISMS

The tool is used to analyze, evaluate and improve business processes. We supplied three kinds of evaluations: predefined, parametrized and user-specific evaluations. The predefined evaluations have been implemented solidly and cannot be modified by the user. The parametrized evaluations are based on given procedures. They may, however, be adapted to the specific application or the respective problem. The user-specific evaluations can be defined by users that have been specifically trained.

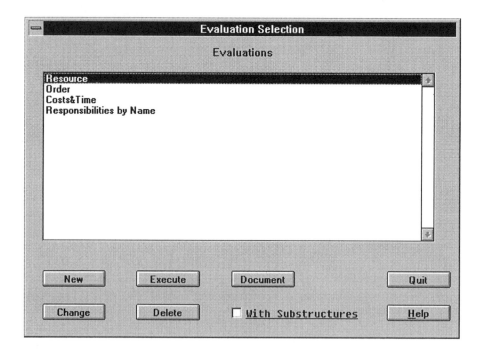

Figure 36: Dialog Window 'Selection of the Evaluation'

7.1 Predefined Evaluations

One goal of the utilization of tools to design business processes is the support of discussions in a project team to improve the processes. For this

Computer-Based Tool 101

purpose, you need report functions that, e.g., generate an index of all elements (terms) of a model automatically or that support the generation of an index in order to find elements easily.

Report functions of this tool enable users to document the information that is represented by the model automatically. Users can choose the desired information that should be documented in a documentation dialog. In particular, you can create an index of the information that is contained in the model.

7.2 Evaluations that can be Parametrized

An essential aspect of defining a new evaluation is the evaluation procedure. The procedure determines how many parameters have to be selected, in which order they have to be selected, whether the parameters require attribute values and of which type these attributes need to be.

To be able to handle these evaluations that can be parametrized easily they are supplied with regard to specific reference classes (Figure 36). Trained users can then create individual evaluations on the basis of evaluation procedures. He needs to select the appropriate evaluation procedure and parametrize it (Figure 37).

7.3 User-Specific Evaluations

With a specific macro language trained users can define new evaluations. The evaluation is carried out through specific attributes of the model elements or through the evaluation that can be parametrized. An example of the application of this type of evaluation is the automatic generation of quality management documents described in chapter 6.

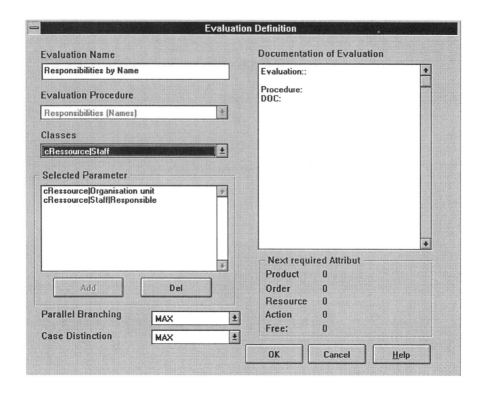

Figure 37: Dialog 'Definition of the Evaluation' following the Selection of all Parameters

7.4 Example of an Evaluation

Figure 38 and Figure 39 show an example of an evaluation of costs and times of a business process. The necessary data is transferred – using a macro – from the model into $MS_®$ EXCEL™. Here, it is summarized in tables and graphically processed.

The hourly rates that are necessary for the calculation have been stored in the resource attributes of the model. The attributes of the processes include the input times with regard to the type of resource (e.g., DP, equipment and facilities, etc.). The process times stored in the model and the costs of the individual processes that were calculated in the tool are then displayed.

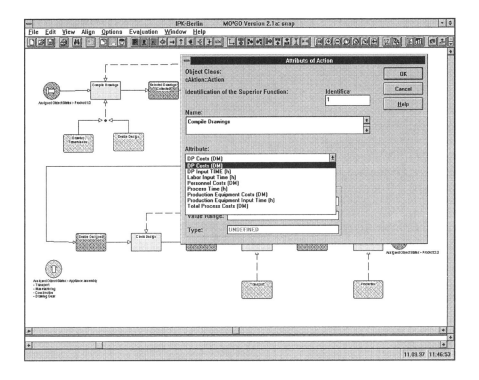

Figure 38: Example of an Evaluation Process

8. SUMMARY AND DIRECTIONS OF DEVELOPMENT

The described tool (MOOGO) to represent, analyze and design corporate structures and business processes allows users to represent and purposefully analyze products, resources, orders and the associated business processes. The advantages of the tool include the systematic planning and optimization process and the reusability of the model for all projects and user views, such as information systems, controlling, quality management and organizational development, that concern the design of business processes.

Reorganization measures and the introduction of new information systems can only be realized sensibly if the users know the existing and the planned business processes. The systematic division of corporate objects into the classes 'product', 'order' and 'resource', that is supported by the tool, results in a transparent representation of business processes and their connections. Refinement functions, modeling rules, reference models or model libraries and consistency tests support a structured approach to modeling (cf. chapters 4. 3, 4. 4 and 5). The opportunities of the object-oriented approach and extensive functions to define objects and business processes enable the user to represent his company-specific concept. The representation in an integrated model is additionally supported by mechanisms for consistency tests, navigation and the modification of models.

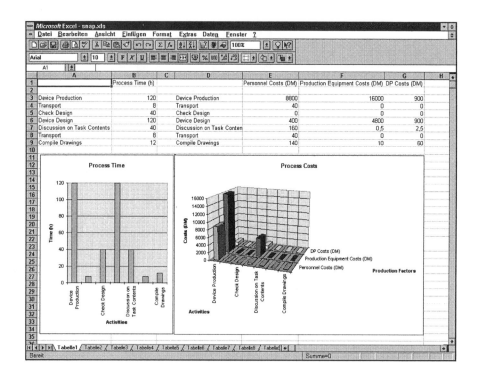

Figure 39: Example of an Evaluation

For the design of business processes, discussion processes within a project team and among different project teams are necessary. For this purpose, the tool provides graphic and text-based documents as a basis for the communication of all participants. The documents contain structured directories of all modeled functions, corporate objects, their documentation and graphics. However, you can also automatically generate documents that conform with the standards according to ISO 9000 ff. from the model. This shortens the corporate certification process considerably.

Computer-Based Tool

The user-specific tool interface of MOOGO enables users to create business process models easily and interactively. The business processes and their connections are represented in different windows in which they can be specified further. The tool also contains mechanisms with which to create the models bottom-up or top-down in any combination. Specific class editors allow you to represent company-specific characteristics of products, orders and resources. The user can also define individual classes or descriptions of features. The description of components is also carried out in the corresponding classes, for example, to create bills of materials. The business processes and their connections can be represented in the respective editors transparently. They can also be specified in any desired way. Navigation and search mechanisms support the orientation within the model.

The tool supports the reusability of submodels and the development of corresponding reference model libraries. There are interfaces with the MS® WINDOWS™ programs WINWORD™, EXCEL™ and ACCESS™.

It is planned to develop the tool into an integrated management system. It is also planned to connect the tool to workflow management systems, cost accounting systems and simulators. The simulation of models is now realized through interfaces with simulation tools. It has already been developed a prototype for the respective connection to a simulator. The connection to mailing systems and the Internet takes the increasing networking of companies into account. The analysis and support of networked companies are other fields of further development.

Chapter 6

Model-Based Development of Quality Management Documents

1. QM SYSTEMS AND QM DOCUMENTS

This chapter describes a method to create quality management documents on the basis of an appropriate model. This model is developed with the modeling language, the reference models and the model libraries of chapter 3, and the modeling rules of chapter 4 and is supported by the tool of chapter 5. The steps of implementation and Certification of a QM System are described. The related case study which describes the application of method and implementation procedure is shown in chapter 7.

For industrial companies the introduction of a QM system and its certification according to DIN EN ISO 9000 ff. are useful in two ways: On the one hand, the introduction of a QM system prompts the development of QM systematics and a control system and the fundamental revision of the organizational structure and the process organization; on the other hand, the certificate inspires confidence in the customers with the quality of the products and is thus often a prerequisite of an awarded order. A certificate according to DIN EN ISO 9000 ff is based on a QM system that conforms with the standards and its documentation in a QM manual. It reflects the fundamental attitude of the management, his intentions and the measures to secure and improve the quality in the company.

The task of the QM manual in the company is to describe the entire QM system [Bau91]. It serves as a constant reference book for all employees. The manual answers the questions who, when, how and with what to conduct quality-assuring measures. It acts as a work instruction for the internal organization of a company and as an orientation for customers, sometimes even for the certification office [Fre93]. In addition, a QM manual must secure an understanding of all quality-relevant activities in the company.

The QM manual describes the quality-specific politics, the specific goals and quality-relevant elements of the organizational structure. It incorporates internal documents, e.g., task descriptions, procedure instructions and work instructions, as well as external rules, such as laws, regulations and guidelines.

The QM procedure instructions are the working basis for the middle management. They describe the methods that are applied in the company along with the predetermined process descriptions, responsibilities, authorities and the necessary documents and resources.

The QM work instruction is the documentation for each individual employee. It can be structured just as the procedure instructions. The amount of included information, however, is much greater.

2. METHODS OF DOCUMENT DEVELOPMENT

2.1 Principle of Element-Oriented QM Document Development

The description of a QM system in smaller or medium-sized companies consists of approximately 150 individual documents. The approach to the development of these documents can be divided into two main methods: an element-oriented method and a model-based method.

The element-oriented approach to develop QM documents is based on the twenty elements of the standard DIN EN ISO 9000 ff.

The twenty QM elements are structured according to process steps. For example, the procurement processes of QM element 6 'procuring', the testing processes of the QM element 10 'testing', the storage processes of the QM element 15 'handling, storing, packing, ...' are summarized and described separately.

An overview of all QM elements and their limits was already given in chapter 3.3.3 'QM Elements as Process Module'.

The element-oriented principle determines that each QM element is described in its chapter. The contents are based on the requirements set in the standard DIN EN ISO 9000 ff.

Model-Based Development of QM Documents

Due to the corporate requirements there are QM procedure instructions for each QM element. In terms of the information included, these are described in much greater detail than in the chapter of the manual. Analogously, there are QM work instructions for the QM procedure instructions that describe the activities of each staff member and the criteria that are required to fulfill these tasks. Figure 40 illustrates the connection between the QM elements with the example of QM element 1 'responsibilities of management'.

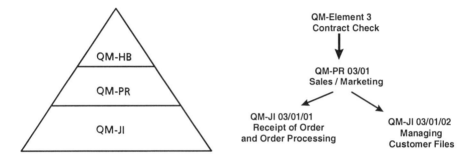

Figure 40: Connection of Chapter of the Manual, QM Procedure Instructions (PR) and QM Job Instructions (JI) – the Example of QM Element 3

Documentation is a costly and therefore controversial task when introducing and maintaining a QM system. On the one hand, the documentation of quality-relevant elements is needed for certification and serves as a job instruction for the employees and thus as basis of process improvements; on the other hand, documentation increases administrative expenses and ties resources and may even interfere with the further development of the company for a longer time. For this purpose, there are several DP programs on the market that should reduce the expenses.

2.2 Program-Technical Support of the Element-Oriented Approach

Many companies use pure word processing programs because they are known and in widespread use throughout the company and because many staff members know how to work with them. Because these programs give users a wide scope the quality of QM documents depends on the professionality of the respective user. The basic structure and the evidential value of the documents can only be ensured if they are continuously examined.

Due to the large number of documents, maintenance is costly. When changing one aspect, many documents have to be adjusted. Insuring consistency and avoiding redundancies of documents requires high maintenance expenses and interferes with the content-wise development of the QM system.

In addition to word processing programs you can also use graphics programs to display the processes transparently. The advantage of these displays is better understanding and higher transparency of the studied aspects. The graphics are used as a basis for discussion when introducing the QM system.

The fact that the interpretation of the displays depends on the person who developed them is a major disadvantage of standard graphics programs. Even though there is a large number of graphics constructs there is no clear descriptive methodology for the aspects. A further disadvantage is that maintenance costs increase because users not only have to maintain and update the texts, but also the graphics. Many users also have to be trained to utilize a graphics program.

To reduce administrative and maintenance expenses drastically, there are database programs that are oriented towards QM documentations and that store QM data without redundancies. Specific QM masks facilitate data entry and data handling. These programs provide interfaces with other standard word processing and graphics programs.

The application of these programs, however, requires additional training expenses. The necessary maintenance expenses to keep the terms in texts and graphics up to date still occur. These database programs are oriented entirely towards QM applications and are thus unfit to be used for other corporate projects.

The structuring criteria of the element-oriented approach are the QM elements whose structure is based on the process steps. The study of processes is, therefore, fairly isolated and concentrates on the requirements of QM.

Time, scheduling and cost aspects are not studied. The integration on the basis of QM structures is – due to the isolated study and the manner of documentation – either connected with excessive adjustment expenses or even impossible. Especially studies of time factors require studies of the entire process. The isolated study of processes does not fulfill this requirement.

2.3 Principle of Model-Based QM Document Development

In contrast to the element-oriented principle the model-based principle studies the processes in a context that includes the entire company. The difference between both concepts thus results from the different goals and structures of the studied processes.

The goal of the model-based principle is to describe the processes in a logical context, i.e., to understand the processes in the sequence in which they actually occur in the company. The element-oriented principle pursues this goal, too. Due to the division of the processes it can, however, be recognized that many processes are studied regardless of the respective context with the company. Examples of this fact include testing and marking processes that occur at different points of the value adding chain.

The model-based principle is based on an enterprise model. The enterprise model is something like a 'map' of the company. This 'map' describes the connection of all processes that focus on value adding. This focus results from the strong emphasis on the marketed product that supplies the company's right to exist. Through the processes the entire product life cycle and the connected management methods and supporting activities are represented. The enterprise model thus combines:
- processes and procedures that determine success with
- the organization, the documents and the resources that are necessary to realize the performance and
- the quality-related indicators, the costs and times that are necessary to evaluate the performance.

2.3.1 Description of QM-Relevant Processes in the Enterprise Model

Due to the different orientation of the processes, the entire process 'business operation' is divided into four subprocesses. These are interconnected and influence each other (Figure 41).

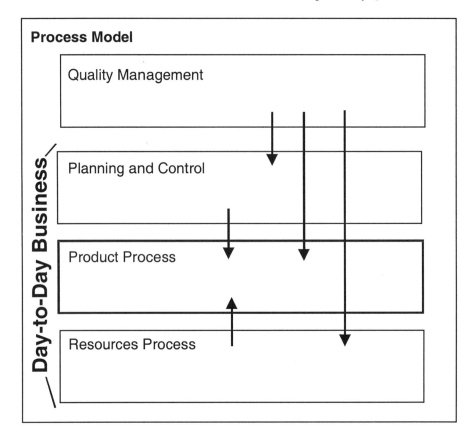

Figure 41: Division of Processes in the Process Model

The subprocesses of *'planning and control'* describe the corporate order handling, beginning with the acquisition of customers and ending with the processing of the customer order – and including all internal and external orders. The entire product life cycle is controlled on the basis of these orders. The product life cycle is described in the *'product process'*. For this purpose, the sequence of the product is represented – beginning with the idea and ending with the maintenance of the product. To support these processes there are supporting processes that are described in the *'resource process'*. Supporting processes include, e.g., processes that describe the handling of testing devices, documents and quality records.

The processes of *'QM'* describe the methods that are used to maintain the QM system. These methods are general procedures that are used on a daily basis, e.g., procedures to conduct internal audits or tests or procedures that take effect if errors occur.

Model-Based Development of QM Documents   113

The process model (Figure 41) contains all processes that are relevant to describe the QM elements. Figure 42 gives an overview of the assignment of QM elements to the process model.

Apart from the processes, the information model is another component of the enterprise model. The information model contains definitions and structures of all objects that are relevant to quality.

2.3.2   Description of Information in the Enterprise Model

The information model defines and describes the structure of objects. These objects include products, orders and resources that should be described in the QM system. The objects are defined and then classified in a so-called class tree. To describe objects, users have to determine features (in IEM called 'attributes'). For products there is, e.g., the feature 'defective'. This feature can have the value 'true' or 'false'. This fact is important to identify and describe the defective products of the product process.

Structuring objects and defining individual features is time-consuming. Therefore, there are prefabricated class trees with associated features specifically for the description of the QM system. Users only need to adjust the class trees company-specifically (Figure 43).

The following chapter describes how to utilize the IEM method to describe a QM manual along with the work instructions and the procedure instructions in the process and information model.

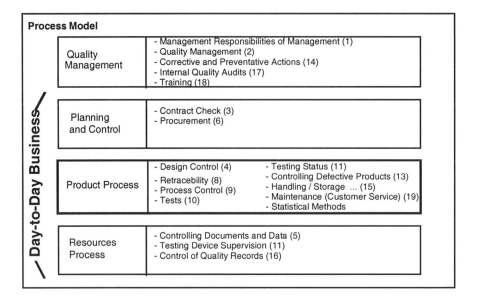

Figure 42: Assignment of the Description of the QM Elements in the Process Model

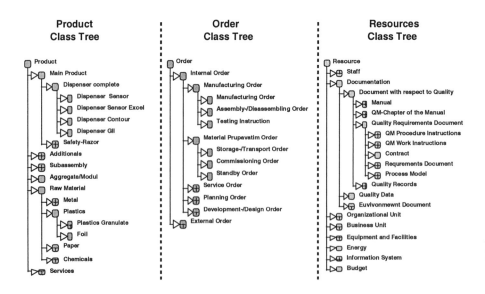

Figure 43: Class Trees to Describe the QM System (Example)

3. DESCRIPTION OF THE QM DOCUMENTS WITH IEM METHOD

3.1 Document Structure

To facilitate the creation of:
- chapters of QM manuals,
- QM procedure instructions (PR) and
- QM work instructions (JI)

the structures of the reference model are used, that were described in the previous chapter.

To be able to identify each QM document clearly is defined as a resource in the class tree (chapter 3.3.2). For example, the document 'QM planning' is assigned to the resource class tree as a subclass of the class 'QM procedure instructions' (Figure 44).

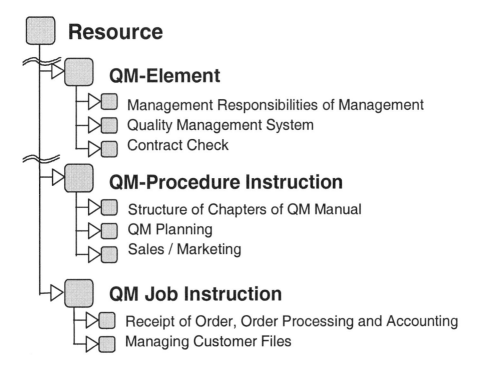

Figure 44: Document Structure Represented in the Resource Structure of the Reference Model

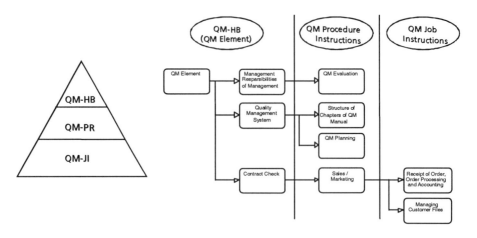

Figure 45: Connection of QM Document Types analogous to the QM Pyramid

The user does not have to file the QM elements of the standard DIN EN ISO 9001 anew: They are already part of the reference class tree. He has to adapt the QM elements according to the standard DIN EN ISO 9002 and 9003, i.e., he has to check which elements are relevant and which elements do not need to be described.

To create the QM document structure, the QM elements, the QM procedure instructions (QM-PR) and the QM work instructions (QM-JI) are defined as resources and are structured with the help of the hierarchy of components (Figure 45). The hierarchy of components of the QM elements corresponds to the table of content of the QM manual.

Each QM element is connected with the associated QM-PR. The QM-PR is connected to the QM-JI (Figure 44). This ensures a continuous connection of all QM documents. The connection is necessary to describe the content of the documents without redundancies. The connection thus ensures that the same content of the QM elements, the QM-PR and the QM-JI only need to be recorded once. The contents include, e.g., the names of the processes, the persons responsible for the processes, documents, facilities and equipment.

3.2 Development of a QM Document

After having described the document structure of the QM system we now go into the individual QM documents. QM documents consist of individual subchapters. Figure 46 illustrates the content-wise structure of the QM-PR and QM-JI. The structure has been taken over from QM documents that have

Model-Based Development of QM Documents 117

already been certified. The contents of the subchapters of the QM element are represented in the attributes of the classes.

> - Chapter Heading (Number and Name)
> - Administrative - Technical Block
> - Purpose
> - Range of Application
> - Responsibilities
> - Definition
> - Process Description
> - Papers that also Apply (not Valid for QM-PR)

Figure 46: Subchapters of QM Procedure Instructions (PR) and QM Job Instructions (JI)

The attributes that are mentioned in the following have been predefined in the reference model. We only use the example of the QM element 15 to explain the content-wise structure because – with only few exceptions – the structure is alike for QM elements, QM-PR and QM-JI.

3.2.1 The Heading of the Chapter

The heading serves to identify the QM element. It consists of a type, a number and a name. The name is identical with the class name of the resource. The type and the number of the document are represented in the attributes 'document type' and 'document number'.

Example of QM Element 15:

Resource Name:	Handling, Storing, Packing and Shipping
Attribute Name:	Document Type
Attribute Type:	List
Attribute Value:	QME

Attribute Name: Document Number

Attribute Type: Text

Attribute Value: 15

3.2.2 The Administrative Block

The administrative block contains administrative data about the document, e.g., the issue number, the date of validity, the name of the person who created the document, the name of the person who checked the document and additional data. These aspects are also represented in the resource class of the QM element as values of the attributes with the same name.

Example of QM Element 15:

Attribute Name: Issue Number

Attribute Type: Integer

Attribute Value: 1

Attribute Name: Date of Validity

Attribute Type: Text

Attribute Value: 1 January 1995

Attribute Name: Created By

Attribute Type: Reference

Class Name: Mr. Sample

Attribute Name: Checked and Released By

Attribute Type: Reference

Class Name: Mr. Sample

3.2.3 The Purpose

The subchapter 'purpose' contains descriptions of the task and the application of the document. To represent this subchapter we use the attribute 'purpose'.

Example of QM Element 15:

Attribute Name: Purpose

Attribute Type: Text

Attribute Value: This QM element contains a description of all measures that are taken to maintain the product quality while handling, storing, packing, preserving and shipping the product...

3.2.4 The Range of Application

This subchapter describes the range of application of the document, i.e., the area the document is relevant for. To describe the range of application the resource class of the document contains the attribute with the same name.

Example of QM Element 15:

Attribute Name: Range of Application

Attribute Type: Text

Attribute Value: This QM element is used in the department 'logistics' and 'operations'.

3.2.5 The Definition

We define terms that contribute to the understanding of the QM element. For the representation, there is the attribute 'definition' in the resource class of the document.

Example of QM Element 15:

Attribute Name: Definition

Attribute Type: Reference

Class Name: Packing (Action Class)

Each term that should be defined has to be defined as an object class. Only then, the term can refer to this object class. The definition is described in the documentation of referred object classes, e.g., in the action class 'packing'.

3.2.6 Co-Valid Documents (Part 1)

Co-valid documents are documentations that are necessary to understand the document. Co-valid documents are represented in the process and in the resource class in the attribute 'co-valid documents' through the hierarchy of components. The attribute 'co-valid documents' refers to documents that cannot be described in the process model. Such documents include, e.g., QM elements or QM-PR that should be read for deeper understanding, that do not however belong to the QM element directly.

Example of QM Element 15:

Attribute Name: Co-Valid Documents

Attribute Type: Reference

Class Name: Correction and Prevention Measures

3.2.7 The Process Description

In this subchapter such processes are explained that are relevant to quality. Process-related descriptions are represented graphically in the process model – through action boxes and associated orders and resources. Each process contains a short description that may be represented in the attribute with the same name in the action box. The process descriptions for the QM elements and the QM-PR may differ in terms of the content of the description. The content of the QM-PR is usually more extensive. Therefore, the process description for the QM elements is represented in the attribute 'short description'; the process description for the QM-PR is represented in the documentation of the action.

3.2.8 The Responsibilities

An important aspect of a functioning QM system is a clear determination of responsibilities (chapter 3.3.2). The processes that are described in the process description thus contain responsibilities and competencies as resources. In Figure 47, e.g., the department head of the store (AL store (v)) is responsible for the process 'packing'. Responsible means: A staff member has to take the responsibility for the result of a process. Competent (or in charge) means: A staff member has to execute the process to yield a result. Positions or business partners may also be responsible or competent. The degree of responsibility is determined in the attribute 'responsibility' in the respective resource. For example, in the resource 'AL store' the attribute 'responsibility' will have the value 'responsible'.

3.2.9 Co-Valid Documents (Part 2)

The processes in the process model not only contain responsibilities, but also documentations. In Figure 47 the list is a document that is needed for packing.

In the previous chapters we described how to represent the content-wise structure of QM elements in the class tree structure and in the process model. To combine the content of the class tree structure and the process model in one document we have to connect the two. In Figure 48 the QM-PR „packing' is connected with the process „packing'.

According to this approach the individual chapters of the QM elements, the QM-PRs and the QM-JIs are represented in the enterprise model. Users can develop a QM view from this model. The QM documents are generated automatically and are then supplied in a word processing program.

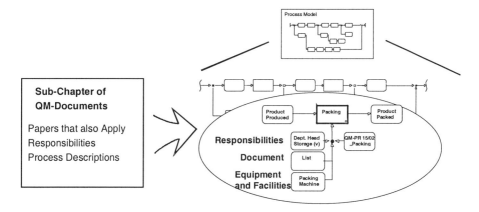

Figure 47: Representation of the Subchapters in the Process Model

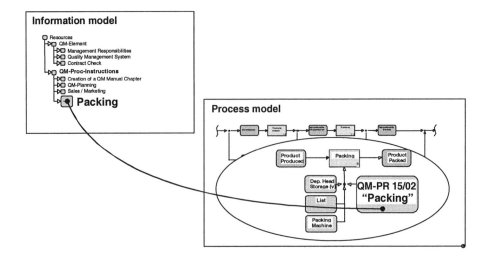

Figure 48: Relationship between Information Model and Process Model

4. AUTOMATIC GENERATION OF QM DOCUMENTS

The creation and maintenance of an enterprise model are supported by the software tool (cf. chapter 5). This facilitates the creation of the model, i.e., the representation of graphic and text-based descriptions in an enterprise model. This program enables you to create formatted QM documents from the enterprise model automatically (Figure 49). These documents include QM manuals that meet the requirements of the standards, QM work instructions and QM procedure instructions.

For development purposes the enterprise model is covered with a QM filter that identifies the content of the QM elements, combines the associated processes and the text-based descriptions and creates a certifiable QM document.

Model-Based Development of QM Documents 123

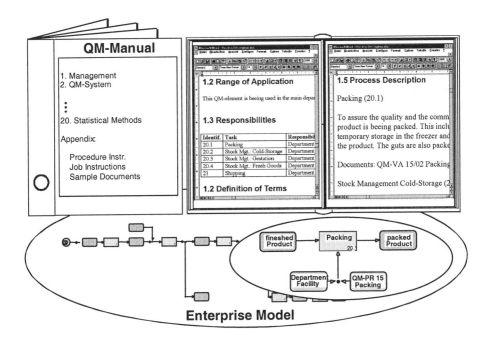

Figure 49: Automatic Creation of QM Documents from the Enterprise Model

5. SUPPORT OF ISO 9000 CERTIFICATION AND IMPLEMENTATION

5.1 Supporting the Certification and Implementation Process

The certification and implementation process according to ISO 9000 is divided into the following steps, and is supported by the following methods (see chapter 3) and tools (see chapter 5).

Step	Methods and Tools Used
Workshop on Goal Definition and Project Delimitation • determine corporate goals and quality policy • resolve relevant standards and corporate processes • establish a project schedule	
Development of an Outline of the QM System • develop the process model up to the procedural level (level 0/1) • structure the QM documentation	IEM/ IEM-Tool (MOOGO)
Initialization of the QM Team • determine the processes to be described for DIN ISO 9000ff • determine the tasks required to develop the QM system • assign tasks to the teams	IEM/ IEM-Tool (MOOGO)
Specify and Implement the QM System • support the stepwise specification of the process model • training sessions and coordination of the development of the QM manual including QM procedural rules, QM job instructions, and checklists • examine the created documents	IEM/ IEM-Tool (MOOGO)
Introduction of the QM System • train the Q manager • train the staff	IEM/ IEM-Tool (MOOGO)
Preparation of Certification • plan and conduct the necessary internal audits • pre-audit as „dress rehearsal" for certification	

Table 3: Steps for QM Implementation and Certfication

The following paragraphs describe the procedure to develop a QM documentation according to ISO 9000ff with the method of IEM, the supplied QM reference model (see chapter 3.3, 3.4.3 and 6.6.2), and the IEM tool (see chapter 5 and chapter 6.1-6.4).

5.1.1 Stage 1: Development of a „Map" of the Company

To get an overview of the corporate value-adding processes, users must initially model the global levels. These global levels are like a „map" that allows all participants to proceed jointly. This stage also ensures that the participants study all processes that are relevant to the QM system.

5.1.2   Stage 2: Determination of the Corporate Goals

The users determine the corporate goals of certification. They have to consider the following aspects:
1. determining the standard according to which the company would like to create its QM system,
2. delimiting the corporate products and processes that are to be examined,
3. setting up a schedule, and
4. developing a plan of measures.

5.1.3   Stage 3: Adaptation of the Manual's Organization

In this stage, the given structure of the QM elements is adapted to the corporate goals. It includes:
1. identifying the QM elements that must not be studied,
2. adapting the chapter-wise structure, and
3. adapting the layout to the corporate design.

5.1.4   Stage 4: Determination of the Procedure Instructions of QM

The user identifies those processes that are relevant to the QM system. He established the procedural instructions (PR) for QM. For this purpose, he proceeds as follows:
1. identifying the processes in the process model that require a QM-PR,
2. defining the QM-PRs in the resource class tree, and
3. connecting the QM-PRs with the appropriate QM element of the manual.

5.1.5   Stage 5: Description of the Contents of the QM Manual

In this stage the user brings the QM elements to life. For this purpose, he describes the processes in the process model, and adds content to the structures in the class tree. The required steps are described in chapters 3.3 and 6.2-6.3. The user should take care to assign all QM-PRs in the process model that belong to an element to the processes.

5.1.6   Stage 6: Description of the Content of the QM-PRs

The user describes the content of the QM-PRs. The following tasks have to be fulfilled:
1. describing the process in the level of detail of the QM-PR,
2. determining the responsibilities for the individual process steps,
3. assigning documents and resources to the process steps, and

4. entering administrative data and text-based descriptions into the designated attributes.

5.1.7   Stage 7:   Determination of the Job Instructions of QM

In this stage the user establishes the job instructions (JI) of QM. He proceeds similarly as concerning the QM-PRs:
1. identifying the processes that require a JI,
2. defining the QM-JI in the resource class tree, and
3. connecting the QM-JI with the QM-PR.

5.1.8   Stage 8:   Description of the QM-JIs

The user now describes the subchapters of the QM-JI in the model. He should in the known manner:
1. specifying the process that is connected to the QM-JI,
2. describing the process steps,
3. assigning responsibilities, documents, and resources (see QM-PR), and
4. entering administrative data and text-based descriptions in the designated attributes.

The development of a QM system is a continuous process. Therefore, steps four through eight are repeated until the QM system is finished. The information that is relevant to QM has thus been portrayed in the enterprise model in a process-oriented manner. The individual QM documents are now generated automatically in text form from this model. For this purpose the information from the process model and the class tree is filtered correspondingly, and then exported into a word processing program. Here, it is formatted (see chapter 6.6).

The advantage is that
- data in the model remains consistent,
- data storage is without redundancies thus limiting maintenance expenses,
- document layout remains the same and formatting occurs automatically,
- data administration, and QM document and responsibilities' administration are handled the enterprise model, and
- process descriptions are available for reengineering purposes.

5.2   ISO 9000 Reference Models

To facilitate and accelerate the development of models regarding the documentation of QM systems, we have supplied a reference model. The reference model consists of pre-structured class trees with attributes for

„products", „orders", and „resources" that are relevant to QM. The attributes allow the user to represent the essential information that is required by DIN ISO 9000ff.

The subsequently designated approach to the description of aspects that are relevant to quality allows users to include the descriptions that are required by DIN ISO 9000ff into the development of an enterprise model (see also chapter 3.4.3).

The reference model can be easily adapted to the conditions of any company. For this purpose, users can rename, move or delete classes according to the specific situation of the company. Users can also add additional classes to the class trees. Only the adoption of the class trees, that have been dealt with in the reference model, allows the automatic generation of QM documents.

5.3 Descriptive Rules of IEM Reference Models

This approach to the description of object classes, processes, and information that is relevant to QM with a reference model relates to the method of „Integrated Enterprise Modeling" that has already been described (see chapter 3).

5.3.1 How are Organizational Units, Documentations, Information Systems, and Resources Described in the Class Tree?

Organizational units, documentations, information systems, and resources are object classes of the class tree „resource". The resource class tree also contains objects – that are classified as classes – that are necessary and able to provide a service (Figure 50).

The user is asked to assign his company-specific classes to the respective reference class tree. The following example of a classification of the object „test protocol" illustrates how users should handle the reference class tree. It is the user's job to find the correct place for the „test protocol" in the class tree, and to assign the object as a new class.

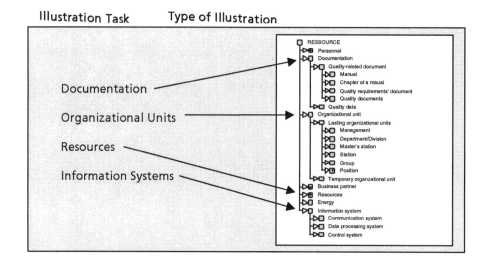

Figure 50: Details of the Resource Class Tree

The following systematic steps must be taken:
1. The „test protocol" is a „resource"!? - **Correct** - proceed to the resource class tree
2. The „test protocol" is a „staff"!? - **Correct** - proceed to the resource class tree
3. The „test protocol" is a „documentation"!? - **Correct** - check the subclasses of the class „document" systematically
4. The „test protocol" is a „quality document"!? - **Correct** - register the „test protocol" as a new subclass of the class „quality documents"

If the structure is too rough for certain applications, the user has to detail the structure further.

The way classes are handled in the process model, i.e., the creation of states and the connection of states to processes, is illustrated in the following figures with examples of organizational units, documentations, and resources.

5.3.2 How are Organizational Units Described in the Process?

The organizational units „project manager" and „Q manager" are examples of the functional description of resources.

The „Q manager" is a position. Therefore, he is defined in the resource class tree as a subclass of the class „positions" (Figure 51).

The process model creates states of „Q manager" and „project manager". With connective elements, the states are connected to the respective activity.

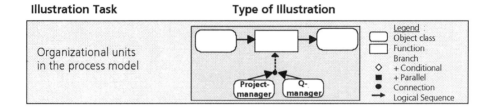

Figure 51: Description of the Resource „Organizational Unit"

5.3.3 How are Used Documents and Required Resources Assigned to the Process Model?

The documents that will be used in a process and the required resources are identified as subclasses of the respective classes „documentations" and „resources" in the resource class tree.

The process model creates states of used „documents" and „resources". With connective elements, the states are connected to the respective activity (Figure 52).

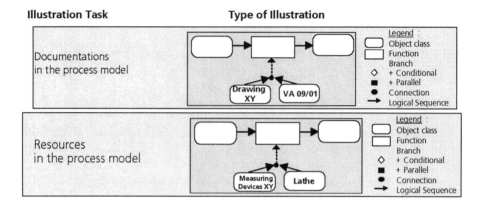

Figure 52: Description of the Resources „Documentation" and „Resources"

5.3.4 How are Changes, for example of Documents, Marked in the Process Model?

The object classes itemized in the reference class trees are specified by attributes that are relevant to the quality. In part, they are given in the class tree. They characterize and delimit the individual product, order, and resource classes precisely. Usually, they correspond to adjectives that describe the object. All objects of an object class are described by the same attributes. They are distinguished by assigning different values to the attributes. The information „changes" may, for example, be reflected by the attribute „status" of an object class. Those objects with the attribute „status" are distinguished by the different values („created" or „changed") of the attributes. By assigning values to attributes, the user describes an object unmistakably.

The following figure illustrates ways to describe changes with attributes that specify objects (Figure 53).

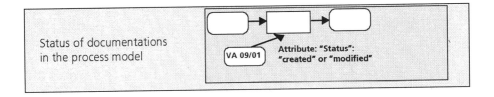

Figure 53: Description of Changes

5.3.5 How are Responsibilities Marked in the Process Model?

Only „positions" or „business partners", such as customers or suppliers, can be „responsible". The degree of responsibility is determined by the value of the attribute „responsibility". For example, the attribute may be specified by the values „responsible" or „competent". „Responsible" signifies that the person is responsible for performing an activity correctly; „competent" signifies that a person has been delegated to perform an activity.

The following figure shows that the „Q manager" is responsible for the given activity (Figure 54).

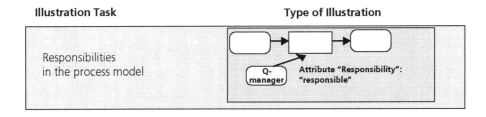

Figure 54: Description of Responsibilities

6. BENEFITS OF MODEL-BASED QM DOCUMENT DEVELOPMENT

The goal of the model-based development of QM documents is the reduction of costs when introducing and maintaining the QM system. Another goal is the reimbursement of funds through the support of reengineering projects.

Promising and cost-saving for the introduction of a QM system are the following aspects:
- the development of the organization of QM is supported by a clear identification of the persons responsible for the processes,
- the description of the processes is easier,
- the QM system only describes those aspects that are really put into action and that can really be proven,
- the expenses required to explain things and to convince employees in the introductory stage are reduced due to the common basis of discussion,
- the QM manual is created from the model in a predefined format and structure.

The expenses for the maintenance of a QM system are divided into maintenance costs of the content, costs for the consolidation and improvement of the system and costs for the updates in the documents.

While carrying out improvement measures, the computer-based enterprise model supports analyses, the detection and evaluation of solutions and the implementation of solutions. The analysis expenses are reduced considerably because once recorded data is available for other projects.

The transparent description of corporate processes facilitates a process- and result-oriented development of the organization. The strong reference to the value adding chains facilitates the identification of cost-intensive bottlenecks in the company. You can develop, analyze and fix different alternative solutions in the enterprise model. A change of the model automatically effects changes of the content of the QM system and thus changes of the QM documents. The time-consuming follow-up documentation is now unnecessary because design and documentation melt into one operation. Modifications only have to be carried out once and effect the entire model. They thus also effect all QM documents.

Apart from the utilization in QM the enterprise model can also serve as a documentation and introduction basis for other management systems. This is possible because management methods, all daily business processes, and indicators concerning time, quality and costs are represented integrally in one model. The tool is able to create different views – for example, as manuals or specifications – automatically from the model.

Chapter 7

Case Study

1. TARGETS AND APPROACH

This chapter illustrates the application of the modeling language (chapter 3), the modeling rules (chapter 4) and the computer-based tool (chapter 5) with a sample project that was carried out at the pilot plant ADITEC (Demonstration Laboratory for Integrated Manufacturing Technology in Aachen). The plant embodies a manufacturing factory that contains most of the systems and the equipment that are used to manufacture goods. As a pilot plant it cooperates closely with research institutes.

In this example we modeled, analyzed and optimized the corporate order handling processes with the help of the tool. The analyses supported by the evaluation functions of the tool focused on identifying weaknesses, developing appropriate measures and modeling the improved target status. Figure 55 illustrates the approach of applying the method and the tool. After that the related QM System was developed and implemented based on the already available model.

To record the actual state we fell back on the available information. We also conducted interviews with staff members. This enabled us to represent the processes in detail and realistically. Details and realism are essential criteria to evaluate to practical applicability.

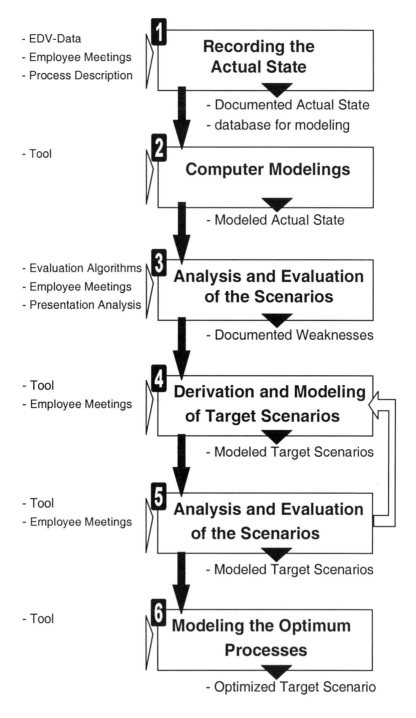

Figure 55: Approach

Case Study 135

The processes were visualized with the help of the IEM-Tool MOOGO (cf. chapter 5). The application of the evaluation opportunities that are implemented in the tool allowed us to identify weaknesses of the processes easily. Further weaknesses, in some cases even combined with appropriate proposals for improvement, were expressed by the staff members when they were interviewed.

The results of these analyses influenced the development of possible target scenarios. These again were analyzed with the tool and with regard to further weaknesses or opportunities to eliminate identified weaknesses. This process resulted in a model that illustrated the optimal target status of the corporate processes. The approach to the development of the model was determined by the available modeling rules (cf. chapter 4).

2. DESCRIPTION OF THE COMPANY

Figure 56 illustrates the most important features of the company. The company manufactures five types of cylindrical gears. The number of units depends on the respective market requirements. Most products are two-phase cylindrical gears that are offered as basic models and as variants. The gears can be employed in different applications (e.g., agitators and printing presses). The company also manufactures customized gears on the basis of available gear types.

To do justice to the numerous applications and to enable an efficient production of a large number of variants the gear is constructed according to a modular principle. The 44 individual parts of the gear were structures into four subunits:
– the power unit,
– the output unit,
– the casing and similar parts' unit and
– the transmission unit.

The casing and similar parts' unit consist of a casting with a cast-iron cover at the transmission and the output member and of the appropriate connecting pieces. Along with the respective flanges the input and output shafts are guided over grooved ball bearings. The are connected to the straight-toothed ball bearings through adjusting springs. A third shaft that is toothed directly and is connected to the additional spur gear through an adjusted spring is guided over grooved ball bearings and a needle-roller bearing. There are additional components for screwing, sealing and oil lubrication purposes.

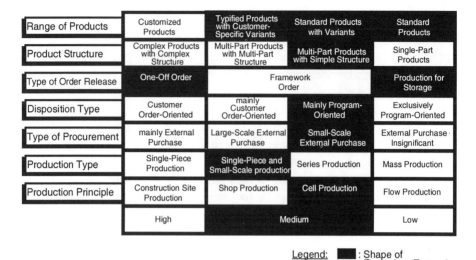

Figure 56: Typology of the Company

The vertical range of production is medium to large because only standardized parts, such as bearing and connection elements, are supplied by external sources.

3. ACTUAL STATE ANALYSIS

3.1 Process Structure of the Company

The studies focused on the order processing system of the company. The processes were modeled and then analyzed with the available evaluation mechanisms. The analyses focused on technological and organizational weaknesses. The identified weaknesses were the basis for the development of appropriate optimization measures. The tool enabled us to represent and evaluate the new target scenarios. Figure 57 illustrates an IEM model of the actual state of the order handling system. According to the IEM method, we selected a description that structures the levels in orders, resources and products. The customer order of the upper level is converted into internal orders that eventually release the product-related activities of the order handling system, such as design, production planning or gear production. The supply, modification or even creation of the resources that are required

Case Study 137

to carry out the activities, e.g., manufacturing or assembly equipment, are represented on the lower resource level. Quality assurance is a cross-sectional task; therefore, it was not assigned to one of the levels, but was represented separately.

To analyze the actual state in detail we specified all essential activities and represented them in the model. The following paragraphs focus on these detailed descriptions of processes (Figure 58). [Mer95f] contains the complete model.

3.1.1   Level 1.1: Offer Preparation

The offer preparation department accepts the inquiries of customers and develops an appropriate offer that is presented to the customer. We distinguish between standardized inquiries that result in standardized offers and non-standard inquiries that require more time and expenses. For non-standard inquiries the company conducts feasibility studies. These might lead to further necessary discussions with the customer.

3.1.2   Level 1.2: Order Handling

The order handling department divides the customer order into internal design, production planning or manufacturing and procurement orders. An order control station decides whether there are basic gear or non-basic gear orders. Non-basic gear orders may be further divided into basic gear batches and non-basic gear batches. Only if there are non-basic gear batches the control station releases design or production planning orders. An early representation of the order and the additionally created data in the PPC system is essential to ensure the integral planning and control of the production.

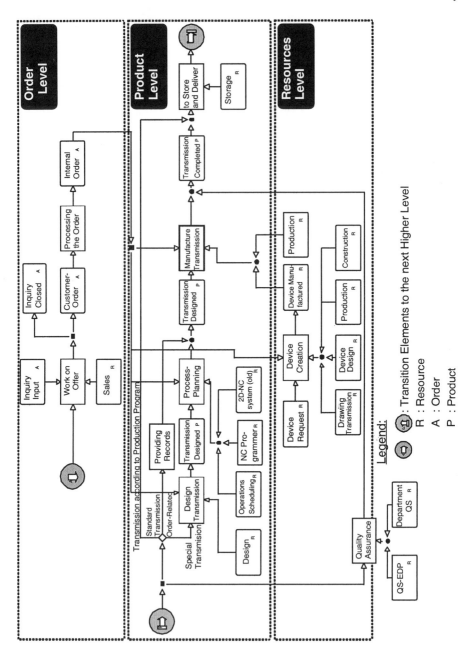

Figure 57: Order Handling System of the Company

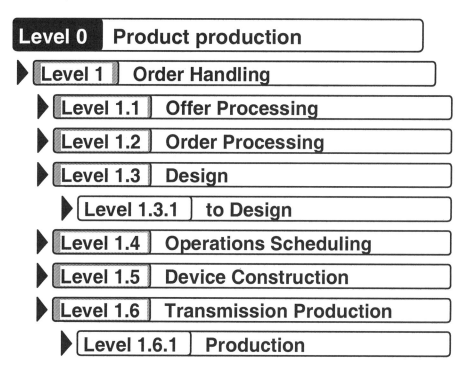

Figure 58: Hierarchical Levels of Order Handling

3.1.3 Level 1.3: Design

The design order releases the production of a new gear. At first, the product is pre-planned, i.e., the company determines technological requirements, e.g., the design specifications and the rough concept. The creation of design documents is represented in the detailed level 'design'.

3.1.4 Level 1.3.1: to design

On the level 'to design' we created the actual detailed design – including the results of drawings and additional documents. With a CAD database we initially study whether existing similar or equal designs of entire gears or components can be reused. If there are no similar designs the components are designed, calculated and worked out. From these individual designs we then create or update the compilation drawing. The respective bills of materials and principal lists of parts are deposited in the PPC system. The head of the design department then checks the consistency, the fulfillment of customer requirements and the feasibility. If corrections then become necessary the design order is redesigned. Finally, the CAD system creates and transfers transfer data for an automatic generation of an NC program.

3.1.5 Level 1.4: Production Planning

Production planning conducts two basic activities. At first, the task and assembly schedule for the entire gear is developed. On the basis of these schedules the required NC programs are designed for the NC machines. The transfer data of design influence this process. The transfer data are represented neutrally. They are adapted to the type of machine within the NC programming process.

3.1.6 Level 1.5: Design and Development of Devices

The company contains an individual department that designs and develops devices that are needed to manufacture gears. The design of the devices is exclusively based on the features of the product; therefore, the devices are designed after the gears have been constructed. This step is strongly effected by available gear drawings which first have to be evaluated. In cooperation with work schedulers, that have scheduled the manufacturing steps, the requirements of the devices are settled. The devices can then be designed correspondingly. After checking the design one last time the devices are manufactured.

3.1.7 Level 1.6: Gear Production

Due to the strong intertwinement, gear production also contains the illustration of materials management, i.e., purchase and coordination of materials – even though this step occurs a little earlier. The activity 'production' includes the production of parts and individual components of the gears. These are then assembled and the entire gear is checked with regard to the required technical specification.

3.1.8 Level 1.6.1: Manufacturing

The submodel 'manufacturing' includes descriptions of the manufacturing processes of the different gear variants.

The modeled processes correspond to the basic order handling processes. Relevant process data, e.g., process times, hourly rates, resource input times and costs, were additionally added.

To support in-depth analyses of production processes the submodel 'manufacturing' was detailed up to the sixth level of the model. The in-depth

Case Study 141

analyses allow users to study identified weaknesses in all their details. Du the large number of data that was deposited users are able to carry out extensive analyses. Especially the application of the DP tool limits the expenses for analyses.

3.2 Description of Time, Cost and Quality Requirements

At the beginning of the project we determined different requirements that concerned the results. These requirements were relevant for all steps of the procedure illustrated in Figure 55. Figure 59 summarizes these requirements and assigns them to the 'traditional' goals of optimization 'time', 'cost' and 'quality'.

A reduction of delivery times was to attain a higher degree of customer satisfaction. This led to the subgoal to break up the existing structures that were oriented towards functions. The goal was to orient the company towards processes.

It was already known that waiting times, particularly during and between production planning and production scheduling activities, were very long. These waiting times, that in some cases lasted several weeks, prolonged the order throughput time considerably and often resulted in an overrunning the arranged delivery dates. Our goal was therefore to study the scheduling departments of the order handling system in detail and to reduce waiting times drastically.

It was also planned to introduce a DP support system that would be technologically and economically sensible. The computer aided systems the company used in production scheduling processes did not always fulfill the requirements of representing and processing the data of the geometrically complex product. In some cases, this resulted in considerable waiting times.

Another problem was that the actual cost drivers within the order handling systems were not transparent. To be able to determine the real reasons for the high manufacturing costs we agreed to assign the gear production costs – as basis of a detailed cost analysis – to the different activities of order handling.

Due to the high costs in the responsible areas and the high expenses for creating and maintaining the manufacturing equipment we concentrated on analyzing and optimizing the manufacturing equipment production process.

To improve the quality of the products and the corporate processes pursued the goal to develop a quality management system according to DIN EN ISO 9001. To induce high synergy effects the introduction was to occur simultaneously with the reorganization of the order handling system. In addition to fulfilling the requirements the goal was to design simple and safe processes. Critical processes, i.e., processes that heavily influence the quality and that in the past have often proven defective, were to be controlled through the integration of control mechanisms. The company expected this measure to result in a reduction of errors.

Time

⇨ Breaking up Functional Structures / Harmonizing and Integration of Processes

⇨ Radical Reduction of Standby Time in the Design Sectors

⇨ Generous Implementation of EDV Support - especially within Production Planning

Costs

⇨ Transparent Illustration of the Actual Cost Producer

⇨ Implementation of a Cost Structure that is continuously Process-Oriented

⇨ Special Reflection on Possible Expense Cuts in the Production of Manufacturing Equipment

Quality

⇨ Realization of the QM Elements according to DIN EN ISO 9001

⇨ Integration of Process Control Mechanisms

⇨ Recording, Processing and Documenting Quality-Relevant Data for Continuos Improvement of the Processes

Figure 59: Requirements of Process Optimization

It was known that only the continuous design and improvement of corporate processes in the sense of CIP (Continuous Improvement Process) could ensure the competitiveness in the long run. Therefore, methods and mechanisms were to be implemented that are able to collect, process and document quality data efficiently and conclusively. This quality database was to enable the company to identify and evaluate weaknesses and to introduce improvement measures.

As explained in chapter 7.1, the method and the appropriate tool were used to analyze and model the processes. The outlined requirements were the decisive basis for analyses and optimization measures. The possibility to define to a large extent the objects and attributes freely allowed the company to represent all attributes in such a way that they fulfilled the requirements. To evaluate the created model the company used the illustration and analysis functions of the tool that were described in chapter 5.

3.3 Development and Description of Weak Points

In the course of the analyses and with the help of the method the company identified several weaknesses and afterwards developed measures to eliminate these errors. Here, we would like to illustrate several central weaknesses. Chapter 7.4 then contains a description of the measures that were taken.

The studied weaknesses were identified in the subprocess chains of design, in production planning and in the design and production of devices. In the following, we describe the weaknesses according to the areas they occurred in.

3.3.1 Design

Concerning the design process the goal was to design the process according to the requirements of the standard DIN EN ISO 9001. The standard's element 'design control' demands that the design data has to be determined, tested and documented. Uncertainties have to be resolved. The design results have to be documented, verified, checked and released. In the stage of utilization the results have to be validated. The company must also determine organizational and technical interfaces of the design process [DIN 9001].

Studies in mechanical engineering have revealed that most errors occur in the planning stages before the actual production starts. These errors are identified much later and thus require much higher expenses [Jah88]. For the design process, this led to the quality requirement that there have to be control mechanisms that eliminate errors at an early stage.

The analysis of the process 'design' of the actual state model revealed that the requirements of the standard have not been fulfilled sufficiently (Figure 60). Until now, design documents have not been determined, tested and documented sufficiently. The design is tested, but there are no measures in effect that could verify and validate the results of the design process. Also, the documentation of design changes has not been organized explicitly. If necessary, the changes are documented; there is no clear rule, however, which staff member is responsible.

Also, there is no procedure that supports avoiding design errors. The resulting disadvantages for the company were quantified by way of an analysis of quality costs for errors originating in design.

The analysis recorded and compared
- error elimination costs (in design),
- testing costs (costs for testing and revising the design) and
- error follow-up costs (rejects' costs, rework costs and warranty costs due to product errors)

for errors that originated in design.

The analysis revealed that every 100 customized orders contained follow-up costs of DM 40.000 that were caused through design errors. Testing costs within design amounted to DM 9.200. The error frequency of customized orders in design was 7%.

Case Study

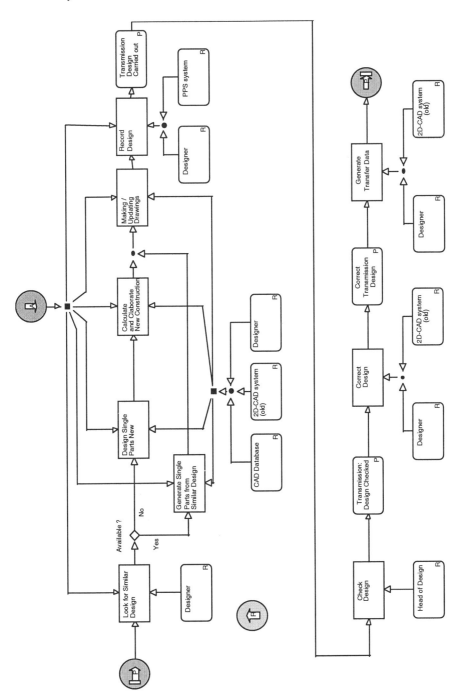

Figure 60: Actual State 'Design'

3.3.2 Production Planning

An analysis of throughput times – that is accelerated and facilitated by the IEM tool – revealed that the throughput time of an order in production planning amounts to five workdays. NC programming alone amounted to three days. A detailed analysis revealed that although the activities have been designed in such a way that they are carried out automatically, manual activities were considerable. This was due to inadequate data exchange processes between the CAD system of design, that also generates the transfer data for NC programming, and the NC programming system.

In the actual state the design department (Figure 61) applied a CAD system that allows users to model the gear that should be designed 2½-dimensionally. Within production planning NC programming (Figure 62) is supported by a two-dimensional NC programming system that uses the transfer data, that is transferred through an interface, of the CAD system to generate the NC program as autonomously as possible.

The detailed analysis revealed that due to the fact that the data of the CAD system is only based on a 2½-dimensional model that is not able to represent the third dimension which is necessary for parts of the gear, e.g., the casing, there are fundamental difficulties in representing the NC model for complex components. The NC programming system was also only designed for two dimensions so that it can only be used efficiently for simple components.

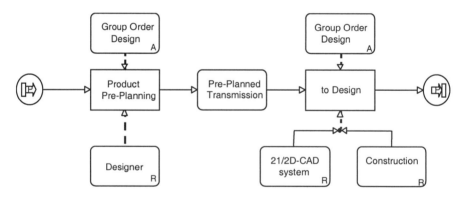

Figure 61: Actual State of Design

This led to the problem that the NC programs had to be redesigned to almost 100% by a process planr who had to be an expert with a high level of qualification. This was highly time-consuming and resulted in high costs.

Case Study 147

The additional personnel input led to process costs in production planning that amounted to DM 820,- per order.

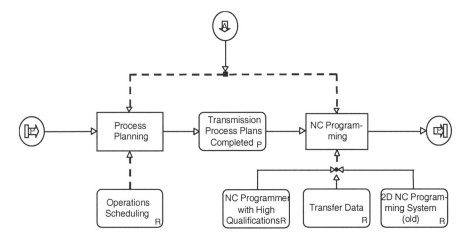

Figure 62: Actual State of NC Programming

Further studies of the design area, that were carried out in the course of this analysis, revealed further difficulties in design that were caused by the 2½-dimensional basis of the CAD system. Volumes of components cannot be calculated and cross-sectional pictures cannot be generated automatically. The system also does not contain a generator that supports the creation of bills of materials. These activities have so far been carried out by the respective official in charge.

In addition, the existing system does not support the search of re-run parts so that documents of components that have already been designed and that can be reused have to be searched for manually. This results in avoidable redesigns. We were not able to estimate the length of these manual activities because the model only records total process times. It does not divide the total time into individual time shares. The opportunity to represent information on data processing systems in the attributes facilitated the analytical processing of the illustrated problems. The illustration and evaluation of the resource input allowed us to determine the comparatively high costs for this resource input.

3.3.3 Design and Production of Devices

The analysis of throughput times with the tool furthermore revealed that there are waiting times of up to two months between the conclusion of production planning and the start of the actual production of the gears. A detailed study led to the realization that this waiting time results from the production of manufacturing devices.

In the actual state model the design of manufacturing devices was an individual organizational area. The organizational connections can easily be made clear in the model through the resource trees.

Due to the differences in measures and processing techniques between the gears the company had to design and manufacture a new device for each new gear (Figure 63). The design of production devices requires – as an input resource – the drawing of the gear that is created in the product design area. It started only if the drawing had been finished. It thus started at the same time as production planning but did not finish at the same time. The time evaluations of the tool enabled us to determine that the throughput time of nine weeks was due to the necessary design and manufacturing expenses about 800% longer.

The gears could only be produced after both processes were finished. The device production department also claimed several resources of the manufacturing department, i.e., of the department that is also responsible for the gear production. It thus reduced the capacity of this department. The resource conflict could easily be illustrated and evaluated in the model.

The throughput time for the design and production of manufacturing devices determined the total throughput time of order handling considerably. It amounted 58%. This led to dissatisfied customers and contributed with DM 2.850 per order considerably to the costs of order handling.

Apart from these cost-producing and rather technical problems within device production there were also quality-oriented weaknesses. The precision of manufacturing devices effects the quality of the product directly. It was therefore one of the goals to reduce the vulnerability of the processes and to take the preciseness of the devices into account when designing the product. For this purpose, the statistical tolerance analysis was named.

An analysis of errors of new designs revealed that 25% of all errors resulted from defective devices. The study of the causes led to the realization that five organizational units (design of production devices, production, design, production planning and production control) participate in designing and producing the devices.

Due to the complexity of the process to design and produce manufacturing devices the number of possible error sources is very high. A sufficient quality of the equipment can thus only be ensured if the expenses are excessive. A further cause was the disregard of device tolerances in the design; the devices were created after the design was finished.

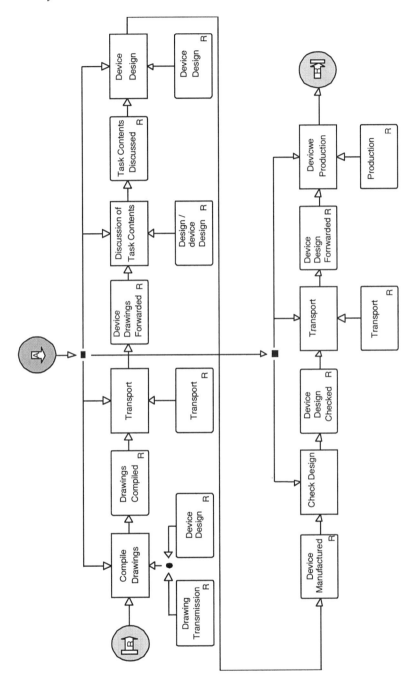

Figure 63: Actual State of the Design and Production of Manufacturing Devices

4. TARGET CONCEPT

Based on the identified weaknesses we developed and evaluated possible scenarios of measures to be taken. The result of these scenarios was a target status of the company that was represented in a target model. The complete model is contained in [Mer95f].

The developed reorganization measures focus on the total production area. In the following, we describe the individual measures according to the weaknesses described in chapter 7.3. Corresponding to the description of the weaknesses the description of the individual measures is again based on the respective corporate areas, i.e., on design, on production scheduling and on the design and production of manufacturing devices.

4.1 Design

The design process was reorganized on the basis of the quality-oriented analysis of weaknesses. To determine and check the design documents in detail we created a checklist. The designers use the checklist before design starts to check whether documents are unclear or even missing. Missing or unclear information is immediately checked with the department that is responsible.

To avoid errors in the design process early on we introduced the Failure Mode and Effects Analysis (FMEA) that follows the development process. FMEA is conducted in interdisciplinary FMEA teams and has the goal to uncover and avoid design errors early on. In the QM system the FMEA process serves to test and verify constructs. Error reports, suggestions for improvements and results of product validation are included. Due to FMEA the drawings only have to be checked once. This reduces expenses considerably.

The introduction of FMEA reduced the design error rate – related to the total number of design orders – from 7% to 5%. For 50 orders the cost of errors that resulted from design now amounted to DM 8.000. For 100 orders these costs would probably amount to DM 16.000. This corresponds to a reduction of costs of 60 %.

The comparison of the reduction of error costs with the reduction of the error rate allows us to say that the introduction of FMEA avoids particularly serious errors.

Figure 64 illustrates the advantages gained through the introduction of FMEA. The reduction of error and testing costs in design is accompanied by an increase of error prevention costs (DM 25.400), that is due to the increase of personnel in FMEA. The financial advantage (DM 3.200) is comparatively low; however, the company also reduced the error rate. A

Case Study 151

lower error rate not only reduces costs and the throughput time (due to less rework), but also increases customer satisfaction.

	Actual Model	**Target Model**
Error Follow-Up Costs	40.000,- DM	16.000,- DM
Testing Costs	9.200,- DM	4.600,- DM
Error Prevention Costs	0,- DM	25.400,- DM
Total Quality Costs	49.200,- DM	46.000,- DM

for 100 Design Orders

Figure 64: Cost Advantages after Introducing FMEA into Design

4.2 Production Planning

After identifying weaknesses in production planning, we developed – with the modeling tool and the evaluation algorithms that refer to the throughput time and process costs – a comprehensive reorganization measure (Figure 65 and Figure 66).

In design, the existing 2½-dimensional CAD system was replaced by a new three-dimensional CAD system. This system enables the full DP support of the mentioned activities. This reduces manual tasks to a minimum.

The NC programming was supported by a new three-dimensional NC programming system that is able to process the transfer data of the new CAD system directly. This reduced the share of manual activities and thus reduced the throughput time. Also, the company could employ an NC programmer with fewer qualifications because the task now primarily focuses on operating the system. Specialized technological knowledge is not required anymore. The highly qualified NC programmer can now be employed in tasks that really suit his skills.

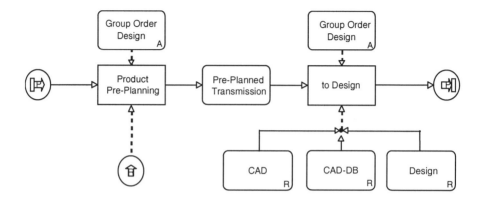

Figure 65: Reorganization of Design

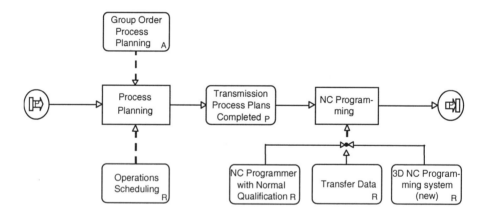

Figure 66: Reorganization of NC Programming

Through the development and evaluation of the target model we were able to determine the savings in the design area. They amount to 16 manhours, which corresponds to a reduction of the throughput time of two workdays (Figure 67). In NC programming the total throughput time was even reduced from 24 manhours, i.e., 3 workdays, to 4 manhours, i.e. half a day.

Case Study

The cost evaluations carried out with the tool revealed saving potentials of DM 1.937 (10%) in the design process and DM 620 (70%) in NC programming.

	Throughput Time		Process Costs	
	Old	Reorganized	Old	Reorganized
Design	198 h	182 h	16489 DM	14552 DM
NC-Programming	24 h	4 h	820 DM	200 DM

Figure 67: Saving Potentials in Design and NC Programming

4.3 Design and Production of Manufacturing Devices

Based on the identified weaknesses regarding a high throughput time, high costs and a high number of product errors, we developed a new concept that focuses on the integration of the design of devices into the production process and on a simplification of all processes. This concept was represented in the target model too. This allowed us to determine the optimization potentials with the tool easily and safely.

One solution was the development of a device kit. This design set contains all necessary standardized components. A device can now be assembled according to the respective design. The production is now limited to the simple assembly of the device. The assembly is based on the requirements that are developed from the design drawings of the gear. The tolerances of the standardized components do not change and are now already taken into consideration while designing the products.

This procedure is incorporated into the gear production process and does not have to be carried out as an autonomous process (Figure 68). The reduction of the organizational units involved leads to a considerable reduction of possible error sources.

154   *Case Study*

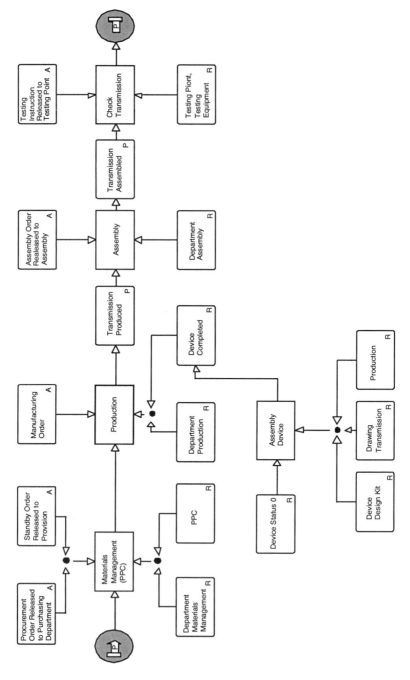

Figure 68: Reorganization of the Design and Production of Manufacturing Devices as an Integrated Component of the Gear Production

Case Study 155

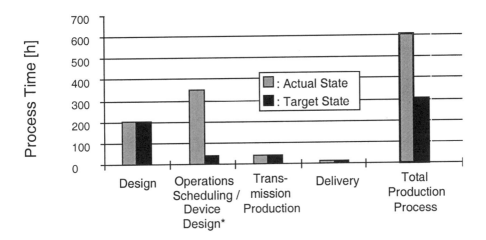

* the Maximum of the Two Parallel Processes

Figure 69: Time-Saving Potentials of the Reorganization Measure

As revealed through the evaluation in the tool, the entire throughput time now amounts to only two hours. It has been reduced drastically. Even short-term changes of customer orders can now be converted quickly.

In terms of the process organization the assembly of devices can still occur at the same time as the production is planned. The entire throughput time of the gears through the production now amounts to 7,5 weeks (up to now: 15,5 weeks). This time now comes closer to the customer requirements. Figure 69 illustrates and summarizes the time-related advantages of the reorganization measures.The investment costs for the device kit amount to DM 25.000. This means that the investment has paid off after only ten gears. This figure was calculated through an investment calculation with the IEM tool (MOOGO). In addition, the process costs of the assembly per gear to be manufactured anew now amount to DM 160. In the actual state these costs amounted to DM 2.850.

The error rate of customized orders was reduced by 76%. After introducing the device kit, only 6% of the errors result from defective devices.

5. IMPLEMENTATION OF QM SYSTEM

5.1 Goals

To safeguard and document the high internal quality requirements, the company planned to develop and certify a quality management system according to DIN EN ISO 9000ff.

The goal was to introduce a standardized quality management system as a chance to further improve internal business processes. On the basis of a model-based description of business processes, QM procedures and methods can be assigned and described clearly.

The tasks were based on the method of Integrated Enterprise Modeling, IEM, supported by the software tool MO^{2}GO and QM reference models from the model library (see chapters 3.3, 3.4.3 and 6.6.2). We worked with a team of the company to support the entire introduction, documentation, and certification process. We consulted, moderated, and trained. The following figure 70 illustrates the goals that were pursued when designing and introducing the QM system.

5.2 Approach

The project was divided into the following steps (see also chapter 6.5.1-6.5.2 and the following figure 71):
1. workshop to define goals and delimit the project,
2. determine the certification association,
3. initialize the QM team,
4. prepare a rough outline of the QM system,
5. consult the company on how to specify and implement the QM system,
6. support the introduction of the QM system,
7. prepare the certification with a pre-audit,
8. training on methods and tools through MO²GO to describe the enterprise model including the QM documents (manual, procedure instructions (PR), and job instructions (JI)).

The modeling and QM software tool MO²GO was used to handle the entire project.

Case Study 157

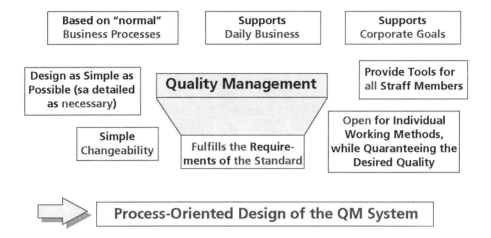

Figure 70: Goals of the Design of the QM System

Approach to the Certification Project

- goal definition, establish relevant standards, schedule, and QM team

- roughly outline the **Process Model**, establish structure of **QM Documentation**

- specify QM system and **Process Model**, create texts on QM elements and PRs, generate QM documentation from the **Model**

- introduce the QM system, supply the QM documentation with the **Models** as aids, training sessions

- certification: selection of the certifier, pre-audit(s), evaluation of the QM documents, certification audit (open)

- stabilization, constant use and update of the QM system and the models

Figure 71: Approach to the Certification Project

5.3 Short Summary of Results

After the project, the company recorded the following results:
- process-oriented QM system had been implemented,
- certificate of the QM system was issued by a well-known certification association,

- QM system was documented in the QM manual, in procedural rules, job instructions, and checklists,
- an enterprise model that contains graphic, (easy to understand) descriptions of business processes, along with used documents and data processing systems (e.g., a database) was created.

In addition, the staff members of the company have been trained in analyzing, autonomously improving, and modeling the processes.

5.4 Target Concept and a Plan of Measures

The development and evaluation of the enterprise model led to the identification of several weaknesses in the corporate business processes. The weaknesses referred to the insufficient fulfillment of standardization requirements, unsatisfactory transparency, inadequate determination of responsibilities, inefficient working methods, and a defective transfer of experiences and applied aids (e.g., checklists). The model of business processes contains a description of target processes, responsibilities, and used documents and data processing systems. All these aspects were discussed with the staff members. This process allowed the company to benefit from experiences, i.e., adopt and „publish" proven, and avoid „defective" processes. The participation of all concerned staff members also ensured that the QM system closely follows the existing processes while simultaneously becoming more efficient. The early information and participation of the staff members ensured a high level of acceptance for the QM system, and reduced the implementation expenses and times drastically.

The target concept was provided with additional checklists and a corporate database. This database allows the company to provide and archive all important project documents for offer and order processing.

The following figure 72 illustrates the structure of the QM manual. The QM system and its documentation are based on the description and structure of the business processes. The company thus implemented a process-oriented system that – in comparison with traditional systems based on QM elements – achieves considerable cost savings in introduction, daily use, and further developments and updates.

The following figure 73 illustrates the structure of the procedural rules, and reveals how this information is described in the enterprise model.

Case Study

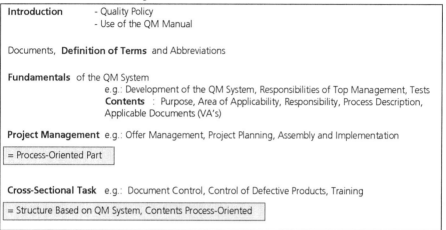

Figure 72: Structure of the QM System

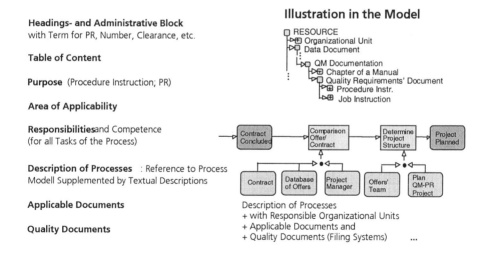

Figure 73: Development and illustration of a Procedure Instruction (PR)

6. FINAL RESULT OF THE CASE STUDY

This chapter contained a description of the practical applicability of the modeling method of 'Integrated Enterprise Modeling' the modeling rules and the supporting tool.

The representation of all data that is relevant to reorganize and optimize is possible. The tool allows users to develop models simply and quickly and to evaluate the relevant model data quickly and purposefully.

The strict modeling rules, e.g., the division into different object classes (product, order and resource), requires a high level of 'modeling discipline'. However, this ensures a high degree of clarity and thus facilitates the analysis of processes. The hierarchical levels of the models allow users to individually determine the desired level of detail. This ensures the transparency of the corporate processes that can be rather complex – particularly in the actual state.

Chapter 8

Standardization

1. INTRODUCTION

In the course of the QCIM project the studies and results regarding Quality-Oriented Enterprise Modeling und the IEM Method (see chapter 3) were presented to the appropriate national, European and international standardization groups that dealt with this topic. The working documents that were introduced are currently either being discussed or are available as proposed standards or pre-standards (cf. List of Standards in annex B). An overview of the concerned national and international standardization groups is presented in Figure 74. In the following chapter we describe into which standardization groups the studies on Quality-Oriented Enterprise Modeling and IEM Method were introduced and which standards were developed or can be expected.

2. NATIONAL

On a national level the results of the studies on Quality-Oriented Enterprise Modeling were introduced into the standardization committee NAM (Normenausschuß Maschinenbau – *Standardization Committee Mechanical Engineering*) 96.5 „Architektur und Kommunikation' (*Architecture and Communication*), into the subcommittee NAM 96.5.1 „Rahmenwerk für eine CIM-Systemintegration' (*Framework for a CIM System Integration*), into NAM 96.4.8 „Industrielles Fertigungsdaten-Management' (*Industrial Manufacturing Data Management*), NAM 96.4.4

„Methodologie, Konformitätstest und Implementierung' (*Methodology, Tests of Conformity and Implementation*), NQSZ-1 „Qualitätsmanagement (*Quality Management*)' and into the standardization committee 'Sachmerkmale' (NSM) (*Product Attributes*).

2.1 NAM 96.5.1 'Framework for a CIM System Integration'

The standardization committee NAM 96.5.1 was initiated by the projects CIM-OSA and KCIM in 1989 and focuses on the standardization of a framework for a CIM system integration. On the basis of the systems theory a framework was developed that contains concepts and rules for the description of companies. This was introduced and discussed in ISO in the mirror committee TC 184/SC 5/WG 1 „Modeling and Architecture' (see below and the list of standards in annex B: ISO 14258: Concepts and Rules for Enterprise Models).

The studies on Quality-Oriented Enterprise Modeling were a basis for the definition of concepts and rules of this framework that should be open to a number of different modeling methods. The studies to develop the framework from the systems theory were based on already conducted studies and results in the QCIM working group 'QUM' [Spu93a]. The conformity of the group's results with the proposed standard was ensured through active participation. A standard with concepts and rules to describe companies will enable a requirements' description of modeling methods, the classification of existing standards in the field of computer-integrated manufacturing (CIM) and the identification of standardization studies that still need to be conducted.

2.2 NAM 96.4.8 'Industrial Manufacturing Data Management'

The standardization committee NAM 96.4.8 was initiated and founded by the KCIM project in 1991. It mirrors ISO TC 184/SC 4/WG 8 (MANDATE; see below) and deals with the standardization of data that are necessary to manage production activities. Through the three generic object classes 'product', 'order' and 'resource' and their attributes Quality-Oriented Enterprise Modeling (QUM) supplied a structuring approach of the data of the project 'Data for Resource Management'. They describe the dynamics of the manufacturing process that is characterized by an assignment of products to resources and by the control of the process with regard to amounts and times. This structuring approach enabled the integration of these views into

Standardization 163

the manufacturing process and reduced the required interfaces between the involved information systems.

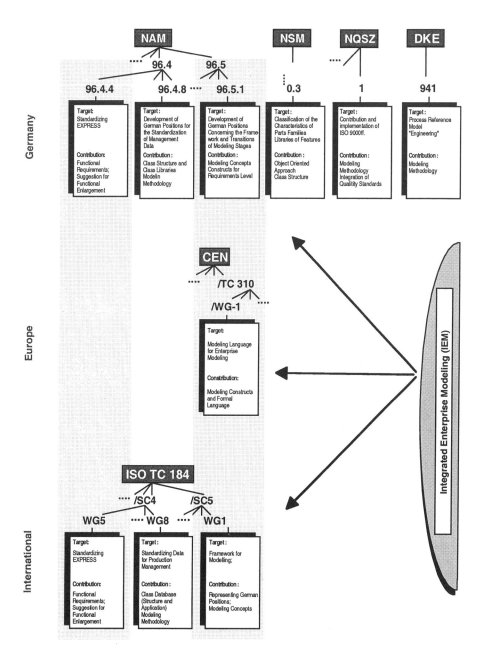

Figure 74: Standardization Activities on Enterprise Modeling

Quality-Oriented Enterprise Modeling was the modeling method that allowed users to determine the necessary data and their relations within the manufacturing process (cf. below and List of Standards in annex B; ISO 15531: Resource Usage Management, Part 1).

2.3 NAM 96.4.4 'Methodology, Tests of Conformity and Implementation'

The standardization committee 96.4.4 worked among other things on the further development and standardization of the information modeling language EXPRESS (cf. chapter 4.2.4).

The working group 'Quality-Oriented Enterprise Modeling (QUM)' supported these studies by phrasing the functional requirements and their conversion in a specification for the object-oriented expansion of EXPRESS. This was to be able to represent functional aspects of describing companies in a formal and standardized language. The specification of the expansion is currently being discussed in international study groups (cf. below and List of Standards in annex B: EXPRESS version 2; Methods/Concurrency Control/ Dynamic Behavior).

2.4 Standardization Committee 'Quality Management, Statistics and Certification Elements (NQSZ) - Subcommittee 1 'Quality Management''

The standardization committee NQSZ-1 focuses on revising the standards of ISO 9000 ff.

The study group 'Quality-Oriented Enterprise Modeling (QUM)' supported these studies by introducing a concept of model-based developments of QM manuals (cf. chapter 7).

2.5 Standardization Committee 'Sachmerkmale' (NSM)

The standardization committee 'Sachmerkmale' focuses on standardized descriptions of products by way of so-called 'Sachmerkmalsleisten'.

The working group 'Quality-Oriented Enterprise Modeling (QUM)' introduced object-oriented modeling as an approach to revise and restructure attributes in class libraries. The generic object classes and their structure of attributes are the basis of an object-oriented structure of the attributes and the possibilities of inheritance rules. When representing relations between the attributes of different object, these relations can be understood clearly.

3. EUROPEAN

On a European level the studies and results of Quality-Oriented Enterprise Modeling were incorporated into the studies of CEN/TC 310/WG 1. In 1990, this committee released a European pre-standard (ENV 40003) of a 'Framework for Enterprise Modelling' that was mainly based on the studies and results of the CIM-OSA project.

Afterwards, the committee developed an 'evaluation report' of modeling methods [CEN91a,CEN91b]. The goal was to find methods that provide modeling constructs for the framework that so far had been empty. The committee analyzed and evaluated 20 methods, including IEM [CEN91]. However, the existing general methods are not adequate and the methods that are currently being developed are not available in sufficient detail. In the course of the QUM project the IEM method was further specified. It was then introduced into the working group as a basis for the development of an appropriate pre-standard (ENV 12204, cf. below).

Further reports (CR 1830-1832, cf. List of Standards in annex B) that were developed concerned the definition of terms and requirements of a corporate infrastructure for the application of enterprise models in the operation and control of business processes and the comparison of corporate infrastructures.

Afterwards, the working group 'QUM' heavily contributed to the development of a European pre-standard 'Modeling Constructs for the Description of Companies (cf. List of standards in annex B: ENV 12204 Constructs for Enterprise Modeling). The active participation ensured the conformity of the modeling constructs of the working group 'QUM' re. ENV 12204.

4. INTERNATIONAL

4.1 ISO/TC 184/SC 5/WG 1 „Modeling and Architecture'

This committee echoes NAM 96.5.1 (see above) on an international level. Since 1989, it works on standardizing concepts and rules for the description of companies (ISO 14258 'Concepts and Rules for Enterprise Models'). The goals are to define a requirement's description of modeling methods, the arrangement and structuring of existing standards in the area of computer-integrated manufacturing (CIM) and the identification of standardization activities that will be necessary in the future (see List of Standards in annex B).

4.2 ISO TC 184/SC 4/WG 5 „EXPRESS Language'

This committee echoes NAM 96.4.4 (see above) on an international level. Among other things, it focuses on the further development and standardization of the information modeling language EXPRESS. The working group 'Quality-Oriented Enterprise Modeling (QUM)' supports these studies with a definition of requirements and an expanded language concept for version 2 of EXPRESS. It should also represent functional aspects of descriptions of companies in a formal standardized language. The expanded language concept is currently being discussed in this working group (cf. List of Standards in annex B: EXPRESS version 2; Methods/Concurrency Control/ Dynamic Behavior).

4.3 ISO/TC 184/SC 4/WG 8 „Manufacturing Management Data' (MANDATE)

This committee echoes NAM 96.4.8 (see above) on an international level. Since 1991, it works on standardizing data for the control and management of manufacturing processes. Especially the relation to and the integration into present STEP activities are taken into consideration. The working group identified three study fields:
- Information that is exchanged with units outside of the studied company.
- Information on resource management.
- Information on the control of the materials flow.

On the basis of the generic object classes, the working group 'QUM' suggested to structure the information in the study field 2 'Data of Resource Management'. QUM can be applied to model the involved business processes and the information that is required to execute the processes. This is the basis for the management and control data that needs to be modeled (cf. List of Standards in annex B: ISO 15531, part 1).

Chapter 9

Summary

Radical organizational changes, i.e., the elimination of complete management levels, have shown that products can leave a company on schedule and with appropriate costs and quality without expensive labor management.

Traditionally, corporate managers try to fulfill the requirements of the market by optimizing individual functions based on the existing organizational structure. This can lead to quite a few interface problems and suboptimal results. A 'cross-functional' study of the integral business processes is often neglected. Different views onto the company and its processes, such as 'quality', 'organization', 'information systems' and 'costs' (controlling), are neither incorporated nor studied nor designed.

Our modelling method (IEM) and the supporting tool (MOOGO) to represent, analyze and optimize corporate structures and business processes enable users to comfortably describe and purposefully analyze products, orders, resources and the respective business processes. The advantages of the tool include the systematization of the planning and optimization process and the reusability of the model for all corporate planning projects and user views, such as regarding information systems, controlling, quality management and organizational development.

To design business processes there have to discussions among the members of a project team and among different project teams. The tool supplies graphic and text-based documents as a basis for communication among the participants. The documents consist of structured directories of all modeled functions, corporate objects, their documentations and graphic descriptions. However, users can also automatically generate documents from the model that conform with the standard DIN EN ISO 9000 ff. This shortens the corporate certification process considerably.

The exchangeability of models between project teams and – in the course of quality requirements of suppliers – between companies is necessary for the general understanding of processes. This requires a uniform modeling language with standardized modeling constructs. Our concept conforms with the standard ISO 14258 „Concepts and Rules for Enterprise Modelling' (cf. List of Standards in annex B). The language constructs are a basis of the European pre-standard for constructs of enterprise modeling (ENV 12204, cf. List of Standards in annex B).

The case study clearly shows that users require a methodical approach and models to carry out reorganization projects. They ensure the common understanding of business processes in the company and enable the users to assign utility potentials of design measures to the respective business processes and resources precisely – as improvements in terms of time, costs or quality. They are thus the basis of any design and optimization process.

Companies are supported to realize the measures of quality-oriented enterprise modeling, to reduce the throughput times, to improve the quality of processes, to reduce costs and thus to improve the orientation towards customers and the competitiveness decisively.

References

[All95] Allweyer, Th.: Modellierung und Gestaltung adaptiver Geschäftsprozesse. Schriftenreihe des Institutes für empirische Wirtschaftsforschung an der Universität des Saarlandes, No. 115, 1995.

[AWK93] Eversheim, W.; König, W.; Pfeifer, T.; et al.: Wettbewerbsfaktor Produktionstechnik - Aachener Werkzeugmaschinenkolloquium, VDI-Verlag, Düsseldorf, 1993.

[Bau91] Bauer, C. O.: Das QM-Handbuch der BAB - Qualitätssicherung. Gießerei Erfahrungs-Austausch 5/91, Fachverlag Gießerei Erfahrungs-Austausch, Heddesheim 1991.

[BT92] Bevilacqua, R. J.; Thornhill, D. E.: Process Modeling. American Programmer, 5/1992.

[CEN91] CEN/CENELEC/AMT/WG-ARC: Süssenguth, W.; Jochem, R.: An object oriented method for integrated enterprise modelling applied for the development of enterprise-related CIM-Strategies and general CIM-Standards. Document N137. 1991.

[Coa91] Coad, P.; Yourdon, E.: Object Oriented Design, Second Edition, Prentice Hall, Englewood Cliffs, NJ, 1991.

[Der94] Dernbach, W.: Geschäftsprozeßoptimierung: „Leidensdruck muß möglichst groß sein...'. Office Management, No. 7-8, 1994.

[DIN8420] DIN/ISO 8420: Qualitätsmanagement und Qualitätssicherung - Begriffe, Beuth Verlag, Berlin, 1992.

[DIN94] DIN EN ISO 9001: Qualitätsmanagementsysteme, Modell zur Qualitätssicherung/QM-Darlegung in Design, Entwicklung, Produktion, Montage und Wartung; edition: August 1994, Beuth Verlag, Berlin.

[DS90] Davenport, Th. H.; Short, J. E.: The New Industrial Engineering, Information Technology and Business Process Redesign. Sloan Management Review, Vol. 32, 1990.

[EK93] Elgass, P.; Krcmar, H.: Computergestützte Geschäftsprozeßplanung. Information Management 1, 1993.

[EM91] Mertens, E.: Der Nutzen von STEP für die CAD /NC-Kopplung, CIM-Management 3/91.

[Eve95]	Eversheim, W.: Prozeßorientierte Unternehmensorganisation, Springer Verlag, Berlin, 1995.
[Fah95]	Fahrwinkel, U.: Methode zur Modellierung und Analyse von Geschäftsprozessen zur Unterstützung des Business Process Reengineering. HNI-Verlagsschriftenreihe, 1995.
[Fer91]	Ferstl, O. K.; Sinz, E. J.: Ein Vorgehensmodell zur Objektmodellierung betrieblicher Informationssysteme im semantischen Objektmodell (SOM). Wirtschaftsinformatik, No. 6, 1991, p. 477ff.
[Fla86]	Flatau, U. (Hrsg.): Digital's CIM-Architecture, Rev. 1.1. Marlboro, MA, U.S.A., Digital Equipment Corporation, 1986.
[Fre93]	Frehr, H.-U.:Total Quality Management „Unternehmensweite Qualitätsverbesserung', Hanser Verlag, Munich, Vienna, 1993.
[Hae92]	Haefner, A.: Untersuchung und Bewertung von rechnerunterstützten Werkzeugen für die Unternehmensmodellierung, Diplomarbeit,TU-Berlin, March 1992.
[Hai93]	Haist, F.; Fromm, H.: „Qualität im Unternehmen - Prinzipien - Methoden - Techniken, Carl Hanser Verlag, Munich, 1993.
[Ham90]	Hammer, M.: Reengineering Work: Dont't Automate, Obliterate. Harvard Business Review, July-August, 1990.
[Ham93]	Hammer, M.; Champy, J.: Reengineering the Corporation, A Manifesto for Business Revolution, Nicholas Brealy Publishing Ltd., London, 1993.
[Ham94]	Hammer, M.; Champy, J.: Business Reengineering, Die Radikalkur für das Unternehmen. Aus dem Engl. von Patricia Künzel, Campus-Verlag, Frankfurt/Main, New York, 2nd edition, 1994.
[HC92]	Harendza, H. B.; Charton-Brockmann, J.: Geschäftsprozesse planen und optimieren. Einsatz von Methoden und Werkzeugen. ZwF 87 (1992), No. 10, Carl Hanser Verlag, Munich.
[HT93]	Hauser, J.; Thurmann, F.: Prozeßmanagement und Systemunterstützung für Concurrent Engineering, CIM Management 2, 1993.
[ISO10303-1]	Industrial automation systems - exchange of product model data. Part 1: Overview and fundamental principles, National Institute of Standards and Technology, USA (1993).
[ISO10303-11]	Industrial automation systems - exchange of product model data. Part 11: The EXPRESS Language Reference Manual. National Institute of Standards and Technology, USA (1992).
[ISO10303-49]	Industrial automation systems - exchange of product model data. Part 49: Process structure and properties. National Institute of Standards and Technology, USA (1992).
[Jah88]	Jahn, W.: Erzeugnisqualität, die logische Folge von Arbeitsqualität, VDI-Zeitung, 130 (1988), p. 4-12; VDI-Verlag Düsseldorf; 1988.
[Joc94]	Mertins, K.; Jochem, R.; Jäkel, F.-W.: Reengineering und Optimierung von Geschäftsprozessen, ZwF 89 (1994) 10, p. 479 - 481, Carl Hanser Verlag, Munich.
[Joc95a]	Mertins, K.; Schwermer, M.; Jochem, R.: Beschleunigung wertschöpfender Prozesse. Ein Beispiel aus dem deutschen Werkzeugmaschinenbau. ZwF 90 (1995) 3, p. 110-112, Carl Hanser Verlag, Munich.

References

[Joc95b] Mertins, K.; Jochem, R.: Business Process Reengineering, Basis for Successful Information System Planning. Winsor, J.; Sivakumar, A. I.; Gay, R. (Editors): Computer Integrated Manufacturing, World Scientific - Singapore, New Jersey, London, Hongkong, 1995.

[Joc95c] Mertins, K.; Jochem, R.: Integrated Enterprise Modelling for Business Process Reengineering. Jansen, H.; Krause, F.-L. (Editors): Life-Cycle Modelling for Innovative Products and Processes, Proceedings of PROLAMAT'95, November 29 - December 1, 1995, Berlin, Germany. Chapman & Hall Publishers. p. 589-600, 1995.

[Joc96] Mertins, K.; Jochem, R.: Modellbasierte Unternehmensplanung - Methode und Werkzeug zur integrierten Modellerstellung. Industrie Management 5/96, GITO-Verlag, 1996.

[Kam90] Kamiske, G. F.: Qualität = Technik + Geisteshaltung. Qualität und Zuverlässigkeit (QZ), No. 5, 1990.

[Kat94] Katzy, B. R.: Unternehmensplanung in produzierenden Unternehmen - Entwicklung einer Methode auf der Basis von Unternehmensmodellen. Shaker Verlag, Aachen, 1994.

[KCIM87] Kommission CIM im DIN (KCIM): Normung von Schnittstellen für die rechnerintegrierte Produktion (CIM). Beuth Verlag, Berlin, 1987 (DIN-Fachbericht 15).

[KCIM89a] Kommission CIM im DIN (KCIM): Schnittstellen der rechnerintegrierten Produktion (CIM), Fortschrittsbericht: Stand der Technik - Ziele in Forschung, Entwicklung und Normung, CAD und NC-Verfahrenskette. Beuth Verlag, Berlin, 1989 (DIN-Fachbericht 20).

[KCIM89b] Kommission CIM im DIN (KCIM): Schnittstellen der rechnerintegrierten Produktion (CIM), Fortschrittsbericht: Stand der Technik - Ziele in Forschung, Entwicklung und Normung, Fertigungssteuerung und Auftragsabwicklung. Beuth Verlag, Berlin, 1989 (DIN-Fachbericht 21).

[Kir96] Kristein, H.: „Denken in Systemen', QZ 1/96, p. 40-42.

[Kle94] Kleinsorge, P.: „Geschäftsprozesse'. Handbuch Qualitätsmanagement, Masing, W., Carl Hanser Verlag, Munich, Vienna, 1994.

[Kos62] Kosiol, E.: Organisation der Unternehmung. Gabler Verlag, Wiesbaden, 1962.

[Loh91] Lohmann, U.: Prozeßkostenrechnung - ein Erfahrungsbericht, Controller Magazin 5, 1991.

[Lüb93] Lübbe, U.: Effizientes Qualitätsmanagement, Unterlagen zum IBM-Kongreß „Industrie und Technik', Garmisch-Partenkirchen, 1993.

[Mer85] Mertins, K.: Steuerung rechnergeführter Fertigungssysteme. Carl Hanser Verlag, Munich, Vienna, 1985.

[Mer91] Mertins, K.; Süssenguth, W.; Jochem, R.: Entwurf und Beschreibung einer Methode zur integrierten Unternehmensmodellierung (IUM), Projektbericht des AK 4.3, Fraunhofer-Institut IPK, Berlin, 1991.

[Mer94] Mertins, K.; Jochem, R.; Hofmann, J.: Referenzmodell Auftragsdurchlauf, Bericht zu AP 5720 im FhG-Projekt „Demonstrationszentrum für Simulation in Produktion und Logistik', Fraunhofer-Institut IPK, Berlin, 1994.

[Mer95a]	Mertins, K.; Jochem, R.; Hermann, J.; Gembrys, S.: Modellbasierte Erstellung eines Qualitätsmanagement-Handbuches. ZwF 90 (1995) No.11, p. 540-543, Carl Hanser Verlag, Munich, 1995.
[Mer95b]	Mertins, K.; Jochem, R.: Qualitätsorientierte Unternehmensmodellierung. DIN-Mitteilungen -Zentralorgan der deutschen Normung, 74 (1995), Beuth Berlag, Berlin, 1995.
[Mer95c]	Mertins, K.; Jochem, R.: Unternehmensmodellierung - Basis für Reengineering und Optimierung von Geschäftsprozessen. „Wirtschaftsinformatik '95', Wettbewerbsfähigkeit, Innovation, Wirtschaftlichkeit, Physica-Verlag, Frankfurt/Main. 1995.
[Mer95d]	Mertins, K.; Edeler, H.; Jochem, R.; Hofmann, J.: Object-oriented Modelling and Analysis of Business Processes. Ladet, P.; Vernadat, F.B. (Editors.): Integrated Manufacturing Systems Engineering, Chapman & Hall, London, Oct. 1995.
[Mer95e]	Mertins, K.; Jochem, R.; Jäkel, F. W.: Tool for Object-Oriented Modelling and Analysis of Business Processes. Proceedings of Joint final conference of ESPRIT Working Groups CIMMOD and CIMDEV, Co-operation in Manufacturing: CIM at Work, TUE - Eindhoven University of Technology, Eindhoven, Netherlands, August 1995.
[Mer95f]	Mertins, K.; Jochem, R.; et al: Qualitätssicherung durch CIM - Teilprojekt Qualitätsorientierte Unternehmensmodellierung. Projektbericht Förderzeitraum 1995, Fraunhofer-Institut-IPK, Berlin, März 1996.
[Mue92]	Müller, St.: Entwicklung einer Methode zur prozeßorientierten Reorganisation der technischen Auftragsabwicklung komplexer Produkte. Dissertation, RWTH Aachen, 1992.
[NN94]	N.N.: Business Reengineering: So erneuern Sie ihr Unternehmen. Gablers Magazin No. 4, 1994.
[Pat82]	Patzak, G.: Systemtechnik - Planung komplexer innovativer Systeme: Grundlagen, Methoden, Techniken. Springer Verlag, Berlin, 1982.
[Pfe93]	Pfeifer, T.: Qualitätsmanagement - Strategien, Methoden, Techniken, Carl Hanser Verlag, Munich, Vienna, 1993.
[REFA93]	REFA: „Methodenlehre der Betriebsorganisation - Lexikon der Betriebsorganisation', Carl Hanser Verlag, Munich, 1993.
[Rie78]	Riehle, H.-G.; u. a.: Systemtechnik in Betrieb und Verwaltung, Teil 1: Grundlagen und Methoden. VDI-Verlag, Düsseldorf, 1978.
[Roh75]	Rohpohl, G.: Systemtechnik - Grundlagen und Anwendung. Carl Hanser Verlag, Munich, 1975.
[Rum91]	Rumbough, J.: Object-Oriented Modeling and Design. Prentice Hall, Englewood Cliffs, NJ, 1991.
[Sce93]	Scheer, A.-W.; Berkau, C.: Wissensbasierte Prozeßkostenrechnung - Baustein für das Lean-Controlling. krp 2, 1993.
[Sch90]	Scheer, A.-W.: CIM-Strategie als Teil der Unternehmensstrategie. Verlag TÜV Rheinland, Cologne, 1990.
[Sche87]	Scheer, A.-W.: CIM - Der computergesteuerte Industriebetrieb, BerSpringer Verlag, Berlin, ^{2}1987 ; ^{3}1988 ; ^{4}1990.
[Sche92]	Scheer, A.-W.; Nüttgens, M.; Keller, G.: Informationsmodelle als Grundlage integrierter Fertigungsarchitekturen. CIM-Management 8 (1992), p. 12-20.

References

[Sci93] Schildbach, T.: Vollkostenrechnung als Orientierungshilfe. DBW 53, No. 3, 1993.

[Scu92] Schuh, G.; Brandstetter, H.; Groos, P.: Grenzen der Prozeßkostenrechnung. Technische Rundschau, No. 23, 1992.

[Sel88] Seliger, G.; Mertins, K.; Süssenguth, W.: Organisation und Planung rechnerintegrierter Betriebsstrukturen CIM. Meins, W. (Editors): Handbuch Fertigungs und Betriebstechnik. Vieweg Verlag, Wiesbaden, 1988, p. 619-653.

[Spu93a] Spur, G.; Mertins, K.; Jochem, R.: Integrierte Unternehmensmodellierung. Beuth Verlag, Berlin, Vienna, Zurich, 1993.

[Spu93b] Spur, G. (Hrsg.): Handbuch der Fertigungstechnik, Vol. 6: Fabrikbetrieb. Carl Hanser Verlag, Munich, 1993.

[Spu94] Mertins, K.; Süssenguth, W.; Jochem, R.: Modellierungsmethoden für rechnerintegrierte Produktionsprozesse. Reihe „Produktionswissen für die Praxis'. Hrsg.: Prof. Dr.-Ing. Dr. h.c.mult. G. Spur. Hanser Verlag, Munich, 1994.

[Süs91] Süssenguth, W.: Methoden zur Planung und Einführung rechnerintegrierter Produktionsprozesse. Carl Hanser Verlag, Munich, 1991 (Reihe Produktionstechnik; Vol. 93). - Gleichzeitig TU Berlin, Fachbereich 11, Diss. 1991.

[TJ92] Tönshoff, H. K.; Jürging, C. P.: CIMOSA - Geschäftsprozeßmodellierung zur Anforderungsbeschreibung für unternehmensspezifische CIM-Anwendungen. CIM Management 6, 1992.

[TöM93] Töpfer, A.; Mehdorn, H.: Total Quality Management - Anforderungen und Umsetzung im Unternehmen. Luchterhand Verlag, Neuwied, 1993.

[Tra90] Traenckner, J.: Entwicklung eines prozeß- und elementorientierten Modells zur Analyse und Gestaltung der technischen Auftragsabwicklung von komplexen Produkten. Dissertation, RWTH Aachen, 1990.

[Vie86] Viehweger, B.: Planung von Fertigungssystemen mit automatisiertem Werkzeugfluß. Carl Hanser Verlag, Munich, 1986 (Reihe Produktionstechnik Berlin; Vol. 50).

[WAL89] Walter, H.-C.: Systementwicklung - Planung, Realisierung und Einführung von EDV-Anwendungssystemen. Verlag TÜV-Rheinland, Cologne, 1989.

[War92] Warnecke, H.-J.: Die Fraktale Fabrik - Revolution der Unternehmenskultur. Springer Verlag, Berlin, Heidelberg, New York, London, Paris, Tokyo, Hong Kong, Barcelona, Budapest, 1992.

[War93] Warnick, B.: Kosten- und Leistungsrechnung als Instrument des Leistungs- und Ressourcencontrollings. krp1, 1993.

[Wie87] Wieneke-Toutaoui, B.: Rechnerunterstütztes Planungssystem zur Auslegung von Fertigungsanlagen. Carl Hanser Verlag, Munich, 1987 (Produktionstechnik Berlin; Vol. 56).

[Wie92] Wiedmayer, G.: Organisatorischer Quantensprung mit Business Process Management. Gablers Magazin No. 8, 1992.

[Wie96] Wiendahl, H. P.: Betriebsorganisation, Carl Hanser Verlag, Munich, 2nd edition, 1986.

[WMC95]   Workflow Management Coalition, Working-Group 1A: Workflow Process Definition Read/Write Interface. Document Number: WFMC-WG01-1000, February 17, 1995.

Appendix A

GLOSSARY

The definitions of the terms were based on the standards listed in annex B. The available definition were adjusted for an application in the project German national project 'QCIM'.

Ability
The amount of all potentially purposeful status information that concern the studied system. The description of the 'ability' contains descriptions of starting states, *transformations* and target states. The current performance describes the processing of given algorithms and transformation rules. The ability to improve the current performance describes the creation of new algorithms and transformation rules.

Abstraction
The activity or the result of disregarding certain differences between objects in order to determine and highlight common features.

Action
Description of the conversion within a transformation.

Activity
Specification of the controlling information through order objects and of the information on executive and processing information through a function.

Actual Characteristic
< is synonymous with the term *'Actual Value'* >

Actual Value
Result of a qualitative or quantitative *feature*.

Arrange Hierarchically
A structuring principle that is characterized by the superordinate/subordinate relations of *units*. Units may be *classes* or *systems*.

Attribute
Description of features of data objects in programs and databases.

Behavior
Acting and reacting of a subsystem or an *element* and the connected modifications of a *system*.

Borders of a System
The interface between system and environment.

Capsulation
Realization of the *principle of secrecy* according to the object-oriented paradigm. *Attributes* of *objects* can only be accessed through *methods* that have been declared as public. The internal structure of *objects* should, however, be invisible for other *objects*.

Characteristic
< is synonymous with the term *'value'* >

Class
An *abstraction* of a number of *objects* that specifies the common features. A *class* is a template with which *objects* can be created.

Community of Interests
A individual or a group of people with mutual interest in the performance of an organization or the environment in which he or they work. As a supplier, each organization has five principle partners of interest: their customers, their staff members, their owners, their subcontractors and society.

Company
Business-oriented legal structure that focuses on achieving lasting profits.

Appendix A: Glossary

Conformity
The fulfillment of certain requirements.

Contractual Principle
The arrangement of a new *object class* into the hierarchy of inheritance signifies that the object class enters into a contract promising that it will also make the methods that are provided by its parent classes available. This ensures that the introduction of a new object class does not change the existing code.

Construct
Symbol of a language that is defined semantically.

Control
General:
Process in which the figure that should be controlled (control figure) is continuously recorded, compared with a guide figure and – depending on the result of the comparison -- adjusted. The resulting process occurs in closed loop, the *control circuit*.
Production:
Authorizing, supervising and securing the execution of tasks with regard to amounts, dates, quality and costs.
Control, II
General:
Process in a *system* in which one or several figures as input figures influence other figures as output figures. This is due to the regularities of the particular *system*. A characteristic of controlling is the open process through individual transfer links or the control chain.
Production:
Prompting the execution of tasks with regard to amounts, dates, *quality* and costs.

Control Circuit
Closed impact pattern of *control*.

Correction Measure
Activity that is carried out to eliminate the cause of an existing error, fault or other undesired situation to prevent the re-occurrence.

Data
Characters that were combined for processing purposes and that – due to known or assumed agreements – represent *information*.

Data Model
Model of describing and processing *data* of an application field and their inter*relations*.

Deviation
Difference between actual and target figures.

Element
Real or imagined things. They are part of a system that are not further divided and that are connected by relations. Elements include, e.g., mountains, rivers, people, companies, machines, schedules, activities, numbers and mathematical symbols.

Entity
Exactly delimited individual specimen of things, persons, terms, etc. The entities are *related*.

Environment
All *elements* that do not belong to the system. In the case of open systems materials, energy and *information* can be exchanged through *interfaces*.

Error
The non-fulfillment of a determined requirement.

Evaluation
A systematic study as to whether a unit fulfills the given quality requirements.

Fault
The non-fulfillment of a desired requirement or a justified expectation of the use of a unit – including safety and security requirements.

Feature
< is synonymous with the term *'Attribute'* >

Formal Language
The language L is the amount of correct sequences of symbols (sentences) that are described by the *syntax*. The *syntax* defines the structure of sentences and encodes a meaning (semantics). A formal language L is characterized by a tupel consisting of four elements (V, N, P, S):

　　V ⇨ Vocabulary of keyboard symbols (symbols that occur in sentences)

N ⇨ Amount of symbols that do not appear on the keyboard
P ⇨ Rules for substitution (Productions)
S ⇨ Starting symbol, an element of N

Formalization
Description of *data* and *functions* with a *formal language*.

Function
A *transformation* in which the input and output either belongs to the object class *'product'*, *'resource'* or *'order'*.

Functional Model
Model of the describing and processing *functions* of an application field and their inter*relations*.

Generalization
The *abstraction* of a number of *classes* that specifies the common features of a parent class.

Goal
States of output figures or states and the behavior desired by the system. Goals can be developed from internal or external requirements and demands.

Hierarchy
A *system* that is characterized by the superordinate/subordinate relations of *elements*.

Information
A message and its meaning. The meaning may be that a person gives meaning to a message. The meaning can also be deduced indirectly from the kind of further processing of the message.

Information Model
Model of the *information* that should be described, processed, stored or created.

Inheritance
Transferring the features of a class to its subclasses. In the subclass these features can be further specified.

Instanciation
Developing an object into an object class.

Interface
Combination of all figures that are required and can be retrieved from the outside of the *system* and general *information* for the application of the *system*. It also includes agreements (protocols) of the manner how *information* should be exchanged

Message
Messages are the only way of communication between *objects*. Messages with *information* can be sent to a certain *object*. The *object* can then react and reply or forward the message.

Method
General:
Consistent method, procedure and activity according to its plan.
Object-Oriented Technology:
A unit of functional logic within an *object*.

Model
Abstraction of a *system*, representing the *elements* and *relations* purposefully.

Modeling Method
Consistent and scheduled methods, procedures and activities to create *models*.

Modeling Tool
Aids that support the application of a *modeling method*.

Non-Conformity
< is synonymous with the term *'Error'* >

Object
Real World:
Phenomenon of the material world that is not necessarily based on consciousness and that is the target of realizations, i.e., something you can describe. Objects of the real world have features that enable the description and the distinction from objects.
Object-Oriented Technology:
Specific features of a *class*, also called an *instance*. The features of objects are represented by *methods* and *attributes*. The static characteristics are described by *attributes*, the dynamic characteristics (*behavior*) by *methods*.

Appendix A: Glossary

Order
Binding entries for *processes* with the *purpose* of coordinating and *controlling processes*.

Organization
On the one hand, the process that summarizes the structure of *relations* and regulations between subsystems and *elements* in a *system*; on the other hand, the result of the described process. The *relations* and regulations should align the *behavior* of the subsystems and *elements* towards a superior *goal*.

Planning
Finding and determining goals systematically, preparing tasks and determining the sequence to attain the goals.

Politics
Long- and medium-term orientation of the management that includes the *goals* and measures to attain the goal.

Polymorphism
In contrast to imperative programming, the object that is called on in object-oriented technology does not have to be known entirely. It is sufficient if the membership of the object to a certain part of the object class hierarchy guarantees the fulfillment of a certain method. The 'call', therefore, is polymorph.

Prevention Measure
Activity that is carried out to eliminate a possible error, defect or other undesired situation in order to prevent the reappearance.

Principle of Secrecy
A modularization principle in the field of software engineering. It indicates that the way a module fulfills its tasks should be concealed in the inside.

Process
A sequence of *transformations* that are interrelated and that transform input into output.
Process, II
A *transformation* that should be planned and controlled. The input of a process are divided into *products*/materials, *resources* and *information*. The

description of a process must not contain the description of *resources* and *information*.

Process Plan
A document that contains the sequence and the description of the processing steps that are required to transfer a workpiece from the starting state into the finished state. The document does not depend on the respective carrier.

Product
Material or insubstantial results of *processes*.

Prototype
Model realization of *systems* (e.g., software, production processes) to verify individual *scenarios*.

Production
All organizational and technical measures that directly serve to convert input materials into intermediate and final products.

Production Equipment and Facilities
Movable and immovable facilities, devices and installations that serve to produce corporate goods and services.

Purpose
Under purposes we understand processes that are carried out by a *system* in its environment or that should be carried out.

Qualification
The status that is awarded to a unit if it has been demonstrated that this unit is suitable to fulfill the given *quality requirements*.

Qualitative Capability
Suitability of an organization or its elements to realize a unit that fulfills the quality requirements of this unit. Elements can be persons, methods or machines.

Quality
The entirety of *features* of a unit regarding the suitability to fulfill predetermined requirements.

Appendix A: Glossary

Quality Check
Determining in what way *objects* or *processes* require the respective *quality requirements*.

Quality Circle
Notional model for the combination of activities that effect the quality of a product or a service in different stages. These stages include the determination of requirements and the final evaluation whether the requirements have been met.

Quality Element
Contribution to quality of a material or insubstantial *product* due to the results of an activity or a *process* in a planning, realization or utilization stage or an activity or a *process* due to an *element* in the process of this activity or this *process*.

Quality Management
All activities of the management task to determine and realize quality politics, goals, measures and responsibilities.

Quality Management - Basic Structure
Basic concept to improve the quality through the combination of prevention methods (feed-forward) and correction methods or measures to treat defective units (feed-back).

Quality Requirement
A statement of the requirements or their conversion into a number of quantitative or qualitative requirements of the features of a unit to allow realization and study.

Relation
Relations structure elements of a system. Examples are the exchange of goods and the currency exchange between countries, static and dynamic relations of place and logical connections.

Resource
Performers that execute or are required to execute *transformations*. Necessary resources for the production of goods include capital funds, staff members, *production equipment and facilities*, information, materials, energy, space and time.

Rule
In computer sciences a rule is a way to represent knowledge. A rule consists of a premise and the conclusion (if ... then ...). If the premise is fulfilled the conclusion can be effected.

Scenario
Closed technical business processes that refer to specific problems and that includes attempts to develop alternative solutions.

Semantics
The study of the meaning of a language.

Specialization
The process and the result of developing subclasses of a class that can be distinguished by additional or further specified characteristics.

Specification
The document that lists the requirements that have to be fulfilled by *objects* or *processes*.

Status
Attributes of the *elements*, their features and inter*relations*.

Strategy
Comprehensive plan to attain goals.

Structure
The abstract framework of *elements* and their *relations* forms the structure of the system. If you only study the type of relation 'materials flow' you arrive at the materials flow structure of the *system*.

Supply Product
The *product* that an organization sells or supplies to customers (the market).

Syntax
Formal structure of sentences and words that belong to a language.

System
An amount of at least two *elements* that interact with each other and the *environment*. The combination of elements into systems is purposeful.

Appendix A: Glossary

Target Characteristic
< is synonymous with the term *'Target Value'* >

Target Value
Given *value* of a qualitative or quantitative *feature*. The actual *values* of this *feature* should deviate as little as possible.

Tool
Aids that support the application of a method.

Total Quality Management (TQM)
Management method of an organization that is based on all members, that concentrates on quality and that focuses on satisfying the customers, on securing success on a long-term basis and on benefiting the members of the organization and society.

Transformation
Transforming input into output.

Treatment of Defective Units
Decision on executive activities and activities to eliminate errors.

Validation
Confirming – through a study and by way of keeping a certificate – that the *quality requirements* of a special application have been fulfilled. In the course of validation it is studied whether the product satisfies the demands of a specific application. The question is, e.g.: 'Did we manufacture the right product?'

Value
Defined unit of the target area.

Verification
Confirming – through a study and by way of keeping a certificate – that the predetermined *quality requirements* have been fulfilled. In the course of verification it is studied whether the product corresponds to the specification. The question is, e.g.: 'Did we manufacture the product correctly?'

View
A selective study of the system that highlights specific aspects and neglects others.

Appendix B

RELEVANT STANDARDS AND OTHER TECHNICAL RULES

In the following, we have listed the standards and the appropriate documents that were dealt with, revised or developed in the course of the integration concept of information management presented in this book. These standards thus reflect the current state of science and technology that is also discussed on an international level. Study papers that contribute to the further development of a standard are highlighted by square brackets that contain the respective standard.

1. DIN V ENV 40003 Rechnerintegrierte Fertigung (CIM); Systemarchitektur; Rahmenwerk für Unternehmensmodellierung / Referenz: 90/270/EWG<M>. 1991-06-00
2. ENV 40003 Rechnerintegrierte Fertigung; Systemarchitektur; Rahmenwerk zur Erstellung von Unternehmensmodellen/Enthält AC Titelkorrektur von 1991/Referenz: 90/270/EWG<M>. 1990-07-00
3. CEN/CLC/R-IT-06 Evaluation Report of Constructs for Views according to ENV 40003. CEN/CENELEC Report (1993-04-00)
4. CR 1830 CIM-Systemaufbau - Wörterverzeichnis. 1995-02-00
5. CR 1831 CIM-Systemaufbau - Unternehmensmodellausführung und Interpretationsdienste - Beurteilungsbericht (Englisch:) CIM systems architecture - Enterprise model execution and integration services - Evaluation report. 1995-02-00
6. CR 1832 CIM-Systemaufbau - Unternehmensmodellausführung und Interpretationsdienste - Festlegung der Anforderungen (Englisch:) CIM

Appendix B: Relevant Standards and Technical Rules

systems architecture - Enterprise model execution and integration services - Statement of requirements. 1995-02-00

7. DIN V ENV 12204 Industrielle Automation und Integration - Systemarchitektur - Konstrukte für die Unternehmensmodellierung; Englische Fassung ENV 12204:1996. 1996-05-00
8. Comparison CIMOSA and IEM Modelling Constructs. CEN/TC 310/WG 1 N41.2 (May 1994)
9. Comparison CIMOSA and IEM Modelling Procedure and Modelling Rules. CEN/TC 310/WG 1 N62 (December 1994)
10. ISO/CD 14258 Industrial Automation Systems - Concept and rules for enterprise models (ISO/TC 184/SC 5 N501; July 1996)
11. Framework for Enterprise Modelling - Version 1.4. ISO/TC 184/SC 5/WG 1 N332 (May 1995)
12. Framework for Enterprise Modelling ANNEX A. ISO/TC 184/SC 5/WG 1 N342.2 (Dec. 1995)
13. Requirements and Methodology for Enterprise Reference Architectures. New Work Item Proposal; ISO/TC 184/SC 5 N 490; May 1996
14. ISO/TR 10314-1 Industrielle Automation; Werkstattfertigung - Teil 1: Referenzmodell zur Normung und Methodologie für die Beschreibung von Anforderungen (1990-12-00; ISO/TC 184; in Überarbeitung)
15. ISO/TR 10314-2 Industrielle Automation; Werkstattfertigung - Teil 2: Anwendung des Referenzmodells auf die Normung und ihre Methodik (1991-06-00; ISO/TC 184; in Überarbeitung)
16. ISO/CD 15531-1 Industrial Manufacturing Management Data. Project 2, Resource Usage Management, Part 1: Overview and fundamental principles (ISO/TC 184/SC 4/WG 8 P2 CD, Kobe, June 1996)
17. ISO/CD 15531-2 Industrial Manufacturing Management Data. Project 2, Resource Usage Management, Part 2: Resource Information Model (ISO/TC 184/SC 4/WG 8 P2)
18. Objective, Structure and Scope of the Information Model for Business Resource Management. ISO/TC 184/SC 4/WG 8 N37.3 Revised Working Draft (Oct. 1995)
19. Industrial Manufacturing Management Data. Project 3, Part 24: Dynamics Systems Conceptual Model (ISO/TC 184/SC 4/WG 8 P3 WD Part 24)
20. Industrial Manufacturing Management Data. Project 3, Part 25: Time Model (ISO/TC 184/SC 4/WG 8 P3 WD Part 25)
21. Industrial Manufacturing Management Data. Project 3, Part 26: Conceptual Model for Modelling Elements (ISO/TC 184/SC 4/WG 8 P3 WD Part 26)

Appendix B: Relevant Standards and Technical Rules

22. Industrial Manufacturing Management Data. Project 3,
    Part 27: Conceptual Model for Flow Modelling
    (ISO/TC 184/SC 4/WG 8 P3 WD Part 27)
23. Industrial Manufacturing Management Data. Project 3,
    Part 28: Conceptual Model for Data Exchange
    (ISO/TC 184/SC 4/WG 8 P3 WD Part 28)
24. ISO 9004 Gliederungsvorschlag für ISO 9004. Beitrag zur Langzeitrevision (März 1994; NQSZ)
25. DIN EN ISO 9004-1 Qualitätsmanagement und Elemente eines Qualitätsmanagementsystems - Teil 1: Leitfaden (ISO 9004-1:1994); Dreisprachige Fassung EN ISO 9004-1:1994. 1994-08-00
26. ISO 9004-4 Technical Corrigendum 1
    Qualitätsmanagement und Elemente eines Qualitätsmanagementsystems - Teil 4: Leitfaden für Qualitätsverbesserung; Korrektur 1. 1994-07-00
27. ISO/DIS 9004-5   Qualitätsmanagement und Elemente eines Qualitätsmanagementsystems - Teil 5: Leitfaden für Qualitätsmanagementpläne. 1994-02-00
28. DIN V ENV ISO 10303-1
    Industrielle Automatisierungssysteme und Integration - Produktdatendarstellung und -austausch - Teil 1: Überblick und grundlegende Prinzipien (ISO 10303-1:1994); Englische Fassung ENV ISO 10303-1:1995. 1996-03-00
29. DIN V ENV ISO 10303-11
    Industrielle Automatisierungssysteme und Integration - Produktdatendarstellung und -austausch - Teil 11: Beschreibungsmethoden: Handbuch der Modellierungssprache EXPRESS (ISO 10303-11:1994); Englische Fassung ENV ISO 10303-11:1995. 1996-03-00
30. ISO 10303-11   Comparison of EXPRESS E2 Extension Proposals. ISO/TC 184/SC 4/WG 5/N240 (August 1995)
31. ISO 10303-11   Requirements for the second edition of EXPRESS. ISO/TC 184/SC 4/WG 5/N252 (May 1996)
32. DIN V ENV ISO 10303-41
    Industrielle Automatisierungssysteme und Integration - Produktdatendarstellung und -austausch - Teil 41: Allgemeine integrierte Ressourcen: Grundlagen zur Produktbeschreibung und -unterstützung (ISO 10303-41:1994); Englische Fassung ENV ISO 10303-41:1995. 1996-03-00
33. DIN V ENV ISO 10303-42
    Industrielle Automatisierungssysteme und Integration - Produktdatendarstellung und -austausch - Teil 42: Allgemeine integrierte

Ressourcen: geometrische und topologische Darstellung (ISO 10303-42:1994); Englische Fassung ENV ISO 10303-42:1995. 1996-03-00
34. DIN V ENV ISO 10303-43
Industrielle Automatisierungssysteme und Integration - Produktdatendarstellung und -austausch - Teil 43: Allgemeine integrierte Ressourcen: Darstellungsstrukturen (ISO 10303-43:1994); Englische Fassung ENV ISO 10303-43:1995. 1996-03-00
35. DIN V ENV ISO 10303-44
Industrielle Automatisierungssysteme und Integration - Produktdatendarstellung und -austausch - Teil 44: Allgemeine integrierte Ressourcen: Konfiguration der Produktstruktur (ISO 10303-44:1994); Englische Fassung ENV ISO 10303-44:1995. 1996-03-00
36. DIN V ENV ISO 10303-46
Industrielle Automatisierungssysteme und Integration - Produktdatendarstellung und -austausch - Teil 46: Allgemeine integrierte Ressourcen: Visuelle Darstellung (ISO 10303-46:1994); Englische Fassung ENV ISO 10303-46:1995. 1996-03-00
37. DIN V ENV ISO 10303-101
Industrielle Automatisierungssysteme und Integration - Produktdatendarstellung und -austausch - Teil 101: Anwendungsbezogene integrierte Ressourcen: Zeichnungswesen (ISO 10303-101:1994); Englische Fassung ENV ISO 10303-101:1995. 1996-03-00
38. DIN V ENV ISO 10303-201
Industrielle Automatisierungssysteme und Integration - Produktdatendarstellung und -austausch - Teil 201: Anwendungsprotokoll: Explizite Zeichnungsdarstellung (ISO 10303-201:1994); Englische Fassung ENV ISO 10303-201:1995. 1996-03-00
39. ISO/DIS 10303-45   Industrielle Automatisierungssysteme und Integration - Produktdatendarstellung und -austausch - Teil 45: Allgemeine integrierte Ressourcen: Materialeigenschaften. 1995-10-00
40. ISO/DIS 10303-47   Industrielle Automatisierungssysteme und Integration - Produktdatendarstellung und -austausch - Teil 47: Allgemeine integrierte Ressourcen: Gestalttoleranzen. 1995-10-00
41. ISO/DIS 10303-49   Industrielle Automatisierungssysteme und Integration - Produktdaten-darstellung und -austausch - Teil 49: Allgemeine integrierte Ressourcen: Struktur und Eigenschaften von Prozessen.1995-10-00
42. ISO/DIS 10303-105 Industrielle Automatisierungssysteme und Integration - Produktdatendarstellung und -austausch - Anwendungsbezogene integrierte Ressourcen: Kinematik. 1995-06-00

Appendix B: Relevant Standards and Technical Rules 191

43. ISO/DIS 10303-202 Industrielle Automatisierungssysteme und Integration - Produktdatendarstellung und -austausch - Teil 202: Anwendungsprotokoll:
Assoziative Zeichnungsdarstellung. 1995-10-00
44. ISO/DIS 10303-104 Industrielle Automatisierungssysteme und Integration - Produktdatendarstellung und -austausch - Teil 104: Finite Elemente Analyse (Entwurf). 1996-05-00
45. ISO 10303-212 Industrielle Automatisierungssysteme und Integration - Produktdatendarstellung und -austausch - Teil 212: Elektrotechnischer Entwurf und Installation (in Vorbereitung)
46. ISO/DIS 10303-213 Industrielle Automatisierungssysteme und Integration - Produktdatendarstellung und -austausch - Teil 213: Anwendungsprotokoll: Prozeßpläne für mittels numerischer Steuerungen zu fertigende Teile.
1996-02-00
47. DIN EN ISO 841 (Entwurf)
Industrielle Automatisierungssysteme - Numerische Steuerung von Maschinen - Koordinatensystem und Bewegungsrichtungen (ISO/DIS 841:1995); Deutsche Fassung prEN ISO 841:1995. 1995-06-00
48. DIN EN 61131-1 Speicherprogrammierbare Steuerungen - Teil 1: Allgemeine Informationen (IEC 1131-1:1992); Deutsche Fassung EN 61131-1:1994. 1994-08-00
49. DIN EN 61131-2*VDE 0411 Teil 500; Speicherprogrammierbare Steuerungen - Teil 2: Betriebsmittelanforderungen und Prüfungen (IEC 1131-2:1992); 1995-05-00
50. DIN EN 61131-2/A11*VDE 0411 Teil 500/A11 (Entwurf) Speicherprogrammierbare Steuerungen - Teil 2: Betriebsmittelanforderungen und Prüfungen; Deutsche Fassung EN 61131-2:1994/prA11: 1995. 1995-12-00
51. DIN EN 61131-3 Speicherprogrammierbare Steuerungen - Teil 3: Programmiersprachen (IEC 1131-3:1993); 1994-08-00.
52. DIN EN 61131 Beiblatt 1
Speicherprogrammierbare Steuerungen - Leitfaden für Anwender (IEC 1131-4:1995). 1996-04-00
53. DIN IEC 65B/236/CD Entwurf IEC 1131-5
Speicherprogrammierbare Steuerungen - Teil 5: Kommunikation (IEC 65B/236/CD:1995). 1995-09-00
54. DKE 951.2 Abbildung der Kommunikationsfunktionen nach IEC 1131-5 und der Sprachelemente der IEC 1131-3 auf die Anwendungsschnittstelle FMS der deutschen Feldbus-Norm DIN 19245-1/2. DKE 951.2

55. DIN IEC 65/196/CDV Funktionsbausteine für industrielle Leitsysteme - Teil 1: Allgemeine Anforderungen (IEC 65/196/CDV:1995). 1995-08-00
56. DIN IEC 65/203/CD Entwurf IEC 1499-2: Funktionsbausteine für industrielle Leitsysteme - Teil 2: Festlegung der Typen von Funktionsbausteinen (IEC 65/203/CD:1995). 1996-03-00
57. DIN EN 61360-1 Genormte Datenelementtypen mit Klassifikationsschema für elektrische Bauteile - Teil 1: Definitionen; Regeln und Methoden
(IEC 1360-1:1995); Deutsche Fassung EN 61360-1:1995. 1996-01-00
58. ISO/IEC 9506-4 Industrielle Automation und Integration; Festlegung der Nachrichtenformate für Fertigungszwecke; Teil 4: Technologiespezifische Ergänzungen für numerische Steuerungen. 1992-12-00
59. ISO/IEC 9506-4 DAM 1
Industrielle Automatisierungssysteme; Festlegung der Datenaustauschformate für Fertigungszwecke; Teil 4: Technologiespezifische Ergänzung für numerische Steuerungen; Ergänzung 1. 1994-03-00
60. Extension of ISO/IEC 9506-4 to include NC Coordinate Measuring Machine Functionality. Working Draft (April 1993; NAM 96.1.3/04-93)
61. DIN 66348-3 Schnittstelle für die serielle Meßdatenübertragung. Teil 3: Anwendungsdienste, Telegramme und Protokolle (Vorschlag für eine Europäische Norm). (1995-04-00)
62. Kurzbeschreibung zu DIN 66348-3 (April 1995)
63. DIN IEC 65(Sec)183 Software-Dokumentation für Prozeßleitsysteme und deren Leistungsmerkmale (IEC 65(Sec)183:1994), (Englisch:) Documentation of software for process control systems and facilities (IEC 65(Sec)183:1994). 1994-10-00
64. Documentation of software for process control systems and facilities (TC 65/WG 7 CD. Jan. 1996)
65. ISO/TR 10562 Industrieroboter - Intermediate Code for Robots (ICR). 1995-05-00

Abbreviations

AP	Application Protocol
API	Application Programming Interface
BMBF	German Federal Ministry of Education, Science, Research and Technology
CAD	Computer Aided Design
CALS	Continuous Acquisition and Life Cycle Support
CAM	Computer Aided Manufacturing
CAP	Computer Aided Planning
CAQ	Computer Aided Quality Assurance
CEN	European Committee for Standardization (Comité Européen de Normalisation)
CENELEC	European Committee for Standardization in Electrical Engineering
CIM	Computer Integrated Manufacturing
CIM-OSA	Computer Integrated Manufacturing - Open Systems Architecture
DIN	German Institute for Standardization
DKE	Committee of German Electrical Engineers in DIN and VDE
EDI	Electronic Data Interchange
EFQM	European Federation for Quality Management
EN	European Standard
ENV	European Pre-Standard
EDIFACT	Electronic Data Interchange for Administration, Commerce and Transport
ESPRIT	European Strategic Programme for Research and Development in Information Technologies

EU	European Union
EXPRESS	Formal Data Specification Language (ISO 10303-11)
ISO	International Organization for Standardization
IT	Information Technology
IEM	Integrated Enterprise Model (IEM)
MS	Microsoft
NAM	Standardization Committee Mechanical Engineering in DIN
NSM	Standardization Committee Product Attributes in DIN
NQSZ	Standardization Committee Quality Management, Statistics and Certification in DIN
OMG	Object Management Group
OMT	Object-Oriented Modeling Technique
PPS	Production Planning and Control (PPC)
prEN	Draft of an EN, available in three languages
QCIM	Quality Management through CIM (BMBF project)
QM	Quality Management
QUM	Quality-Oriented Enterprise Modeling
SC	Subcommittee
SCADA	System Control and Data Acquisition
SQL	Structured Query Language
STEP	Standard for the Exchange of Product Model Data (ISO 10303)
TC	Technical Committee
WG	Working Group

Index

activity 9, 14, 21, 22, 24, 26, 28, 34, 35, 40, 41, 47, 63, 78, 79, 80, 86, 129, 131, 140, 175, 177, 180, 181, 183
attribute 3, 18, 24, 30, 32, 33, 36, 38, 52, 53, 54, 56, 58, 71, 75, 78, 80, 81, 90, 92, 96, 98, 101, 102, 113, 117, 120, 130, 143, 162, 164, 176
business process 5, 6, 7, 8, 10, 13, 15, 27, 36, 37, 50, 59, 68, 75, 93, 94, 99, 100, 102, 103, 104, 105, 132, 156, 158, 165, 166, 167
  design 8, 11, 14, 17, 89
  model 7, 13, 17, 24, 25, 89, 90, 98, 105
  modeling 89
certification 37, 38, 67, 104, 108, 109, 124, 125, 156, 157, 164, 167
class 18, 22, 24, 32, 35, 42, 55, 61, 72, 75, 78, 83, 90, 91, 93, 96, 98, 99, 100, 105, 128, 164, 176, 177, 179, 181, 184
  tree 38, 62, 63, 77, 113, 116, 121, 125, 126, 128, 130
construct 3, 5, 22, 27, 66, 74, 110, 150, 165, 168, 177, 187
controlling 8, 9, 11, 20, 22, 45, 46, 47, 103, 167
cost 7, 8, 9, 14, 37, 48, 50, 65, 79, 89, 90, 102, 105, 110, 132, 140, 144, 146, 147, 148, 150, 153, 155, 158, 167, 177

document 21, 36, 37, 38, 40, 44, 45, 48, 66, 67, 108, 117, 119, 126, 142, 182, 184
documentation xi, 37, 40, 46, 66, 85, 101, 104, 107, 108, 120, 121, 124, 126, 129, 132, 143, 156, 158, 167
element 1, 3, 6, 9, 24, 29, 33, 36, 37, 39, 44, 45, 61, 62, 66, 69, 77, 84, 96, 98, 101, 108, 111, 113, 116, 125, 129, 143, 158, 164, 176, 179, 181, 182, 184, 188
enterprise model 17, 18, 36, 37, 44, 58, 59, 61, 62, 66, 85, 90, 111, 122, 126, 127, 132, 156, 158
Enterprise Modeling 90, 97, 127, 156, 160, 162
generic activity model 22, 26
IEM method 30, 62, 77, 90, 95, 113, 136, 165
  model 38, 45, 47, 83, 91, 136
  object 18, 32, 34, 82
information model 24, 38, 68, 74, 83, 93, 113, 164, 179, 188
information system 3, 6, 8, 9, 13, 50, 68, 103, 163, 167
job instruction 109, 117, 123, 124, 156, 158
model library 17, 49, 70, 77, 80, 156
modelling language 15, 17, 41, 89, 93, 107, 133, 164, 166, 168

196　　　　　　　　　　　　　　　　　　　　　　　　　　　　　　　*Index*

method 3, 6, 24, 61, 65, 89, 160, 162, 164, 165, 180
rule 15, 49, 50, 66, 67, 69, 104, 107, 133, 135, 160
tool 14, 69, 89, 151
object 1, 6, 18, 20, 21, 24, 26, 27, 31, 32, 33, 34, 35, 38, 44, 49, 55, 61, 62, 63, 68, 71, 72, 73, 74, 75, 77, 78, 80, 81, 82, 84, 90, 91, 92, 95, 96, 103, 104, 113, 120, 127, 130, 143, 160, 162, 164, 166, 167, 175, 176, 177, 179, 180, 183
-oriented 17, 24, 62, 74, 90, 92, 104, 164, 176
order 20, 22, 24, 27, 36, 42, 44, 45, 48, 49, 50, 53, 54, 55, 56, 61, 69, 78, 81, 101, 112, 136, 139, 140, 141, 146, 148, 160, 175, 181
organization 6, 7, 9, 11, 12, 15, 21, 37, 38, 39, 40, 42, 44, 47, 49, 50, 53, 58, 62, 67, 68, 73, 75, 85, 86, 89, 97, 107, 108, 111, 132, 143, 148, 153, 167, 176, 182, 184, 185
procedure instruction 37, 39, 63, 78, 108, 109, 115, 116, 122, 156
product 11, 20, 37, 44, 45, 48, 50, 59, 78, 87, 93, 97, 111, 112, 113, 119, 139, 144, 150, 170, 183, 184, 185

production 2, 10, 11, 14, 22, 37, 44, 55, 59, 63, 85, 135, 136, 139, 141, 143, 150, 153, 155, 183
QM system 38, 43, 45, 61, 63, 66, 77, 84, 110, 112, 116, 126, 132, 150, 158
quality management 9, 11, 17, 37, 43, 45, 46, 49, 62, 63, 64, 65, 66, 67, 69, 73, 77, 79, 80, 82, 84, 85, 98, 101, 103, 156, 167, 183, 185
reference model 17, 43, 50, 56, 64, 68, 69, 77, 79, 86, 89, 104, 105, 107, 115, 117, 124, 126
relation 1, 3, 4, 6, 9, 18, 20, 21, 22, 23, 33, 38, 40, 48, 90, 96, 99, 164, 166, 176, 183
reorganization 86, 87, 141, 150, 152, 154, 168
resource 8, 21, 22, 27, 36, 38, 40, 42, 50, 53, 55, 63, 78, 102, 112, 115, 118, 119, 120, 121, 125, 126, 127, 128, 129, 130, 137, 140, 147, 148, 162, 164, 166, 188
system 1, 3, 6, 9, 13, 24, 38, 40, 44, 61, 62, 67, 70, 73, 74, 80, 82, 84, 85, 93, 105, 109, 116, 120, 126, 128, 132, 136, 137, 140, 151, 156, 158, 162, 176, 178, 184, 185
TQM 7, 12, 185
workflow 93, 105

PUBLISHER DISCLAIMER

This CD-ROM is distributed by Kluwer Academic Publishers with *ABSOLUTELY NO SUPPORT* and *NO WARRANTY* from Kluwer Academic Publishers.

Use or reproduction of the information provided on this CD-ROM for commercial gain is strictly prohibited.

Kluwer Academic Publishers shall not be liable for damage in connection with, or arising out of, the furnishing, performance, or use of this CD-ROM.

For questions about the contents of the book and the usage of the disk in the back cover, please contact:

Roland Jochem
Fraunhofer Institute for Production Systems and Design Technology
Department Systems Planning
Pascalstrasse 8-9
10587 Berlin
Tel: +49-30-39006-195
Fax: +49-30-3932503
E-mail: roland.jochem@ipk.fhg.de
Internet: http://www-plt.ipk.fhg.de